N·O·R·M

A GUIDE TO NATURALLY OCCURRING

RADIOACTIVE MATERIAL

N·O·R·M

A GUIDE TO NATURALLY OCCURRING
RADIOACTIVE MATERIAL

WILLIAM FEATHERGAIL WILSON

PennWell Books

PennWell Publishing Company
Tulsa, Oklahoma

Copyright © 1994 by

PennWell Publishing Company

1421 South Sheridan

P.O. Box 1260

Tulsa, Oklahoma 74101

Wilson, William Feathergail.

A Guide to Naturally Occurring Radioactive Material (NORM)

Library of Congress Cataloging-in-Publication Data

Wilson, William Feathergail.

    A guide to naturally occurring radioactive material (NORM) /

William Feathergail Wilson.

        p. cm.

    Includes bibliographical references and index.

    ISBN 0-87814-407-2

    1. Radioactive substances – Health aspects. I. Title.

RA569.W53   1993

616.9'897–dc20

Printed in the United States of America

1 2 3 4 5   97 96 95 94 93

CONTENTS

PREFACE vii

CHAPTER 1 • INTRODUCTION AND DEFINITIONS 1

CHAPTER 2 • BASIC RADIOACTIVITY THEORY AND CONCEPTS 11

CHAPTER 3 • CONSTITUENTS OF NORM 39

CHAPTER 4 • GEOLOGY OF NORM 49

CHAPTER 5 • NORM RADIATION UNITS 67

CHAPTER 6 • INDUSTRIAL SOURCES OF NORM 81

CHAPTER 7 • HEALTH RISKS ASSOCIATED WITH NORM 91

CHAPTER 8 • OVERVIEW OF NORM REGULATIONS 123

CHAPTER 9 • RADIATION PROTECTION 139

CHAPTER 10 • RADIATION INSTRUMENTATION 155

CHAPTER 11 • SURVEYS AND SAMPLING 163

BIBLIOGRAPHY • NORM LITERATURE 183

APPENDIX • ACRONYMS, SUMMARY FACTS, REPORTS, AND DEFINITIONS 195

INDEX 209

PREFACE

This book was originally written as a teaching manual for a NORM short course to be given as a one-day seminar. The book grew from this seed, which was planted in 1992. The loss of an eye lens to radiation provided a rather compelling reason to begin research on the subject of NORM. The author used to work in the uranium fields of South Texas, as a young geologist in graduate school at the University of Texas in Austin, during the early 1960s. The examination of uranium ore with a hand lens set the stage for latent damage.

The subject of NORM is certain to change with time as knowledge and research expands. The regulations governing NORM are just now beginning to grow within the states and the federal regulatory domain as this first edition is completed. It is hoped that this book will serve as a proactive aid and introductory manual to those workers that will follow. The book was written at a level of understanding to encompass all persons interested in the subject of NORM. The author welcomes suggestions and recommendations for future editions. It is an early stage for NORM study, and the future will bring forward many new studies and research efforts by a varied group of workers. Hopefully, this book will serve as a platform for knowledge expansion, and perhaps it will trigger a few new ideas on how to deal with NORM.

ACKNOWLEDGEMENTS

The author wishes to thank the publisher, PennWell, for its support and help with launching this first edition on NORM. He also wishes to thank his two sons, Douglas Hord Wilson, geologist, and Clayton Hill Wilson, geologist, who both added suggestions and editing skills to the final copy. A special gratitude is reserved for the author's half-brother, Robert Wilson, attorney, McGinnis, Lochridge, & Kilgore, for his timely reports on the development of NORM regulations in the state of Texas. Finally, the author wishes to thank Elizabeth Gail Wilson, research librarian, Deloitte & Touche, for her support as a spouse, editor, and most of the data base research efforts that went into the book.

William Feathergail Wilson

INTRODUCTION AND DEFINITIONS

INTRODUCTION

Naturally Occurring Radioactive Material (NORM) is a widespread substance in the earth's environment. NORM exists in soil, water, plants, petroleum, coal, lignite, phosphate, geothermal waste, waste water, humans, and animals. The human body cannot directly sense or detect NORM through any of its sensory mechanisms. A false sense of security may exist in a NORM situation, which may present an actual or latent threat to the health and safety of an individual. NORM can only be detected and measured indirectly with specialized instrumentation.

NORM regulations have been declared in several states and are in various stages of development in many more states. The federal government is also studying the subject, and it is believed that the *U.S. Environmental Protection Agency (EPA)* will also adopt federal regulations concerning the treatment, storage, and disposal of NORM within the next few years.

Radionuclides are specific types of atoms that have measurable "lives," usually denoted as *half-lives.* They are radioactive due to a transformation process from one state of energy and matter to another, in a specified period of time. The conversion process, or *decay,* sheds radioactive particles, known as *alpha* and *beta* particles, accompanied by *gamma* radiation. NORM radiation can be damaging to humans, both on a short-term and a long-term basis, if the radiation is concentrated. These topics will be explored along with a few definitions in the initial paragraphs of this book to allow the reader an introduction into the realm of NORM.

The primary NORM *radionuclides* include the elements *uranium, potassium, radium,* and *radon.* NORM waste falls under the category of *Technology Enhanced Natural Radioactive Material (TENR).* TENR is the acronym used to describe the NORM waste resulting from various industrial activities that alter and concentrate the physical state of NORM. For example, scale may build up on the inside of oil well production tubing and may concentrate considerable quantities of radioactive material that has the potential to expose humans to relatively high doses of radioactivity. Additional pathways of human exposure emanate from phosphate mining, coal ash electric generating stations, and ground water introduced into public and private water systems.

NORM is often precipitated as *sludges, scales,* and *residues* in waste streams. Radon gas is sometimes present as a natural element in ground water, as well as soil. Radon and radium also occur in oil and gas reservoirs with associated brine. Several freshwater aquifers in the Houston metropolitan area are loaded with radon radioactive gas. A few Houstonians can thus refer to their baths and showers as "glowing." Organic material such as oil, coal, and lignite chemically "soaks" up NORM radioactive elements and concentrates the material. Soft coal and lignite are especially noted for their affinity for radioactive minerals. Natural gas is much less of a NORM hazard than coal or lignite, but the nation's energy balance for electrical generation is tipped toward coal and lignite, due to monetized special interest groups rather than environmental considerations and price. The public is almost totally unaware of the scales and sources of environmental risk.

In 1991, for example, a Louisiana biologist reported that waste water from an oilfield near Bayou Terrebonne was discharging radioactive NORM waste at levels 20 times higher than allowed discharge levels at the Waterford nuclear power plant near Taft, Louisiana. This plant discharges waste into the Mississippi River.[7]

Regulations are now rapidly evolving in several states, as mentioned previously. The position of NORM regulations as of March 1993 are outlined below:[1]

▼ **NORM — Regulating states**

Louisiana

Mississippi

Arkansas

Texas

▼ **NORM regulations expected in 1993**

Alabama

Kentucky

New Mexico

Oklahoma

▼ **NORM regulations in preparation**

Illinois

Michigan

▼ **NORM regulations anticipated**

Alaska

Kansas

Ohio

▼ **NORM states waiting CRCPD guidelines**

California

Colorado

North Dakota

Pennsylvania

▼ **NORM states that believe they are covered adequately by existing regulations**

Nebraska

Nevada

Utah

West Virginia

▼ **NORM states that have no present action plans**

Florida

Indiana

Montana

South Dakota

Tennessee

Wyoming

A few states without oil and gas production-related NORM problems are also considering regulations. They include New Jersey and Washington.

NORM was first recognized as a potential problem as long ago as 1904 in the oil fields of Canada.[2] It became an issue in the North Sea oil and gas production facilities in the early 1980s[3] and became more widely recognized in the United States in 1986 in the state of Mississippi during a routine well workover. The story was highly publicized in headlines and feature articles.[3] It was not uncommon for goal posts in Louisiana soccer fields to be constructed from oilfield tubing. It was subsequently discovered that a few of these goal posts were radioactive. Thus it was not happenstance that the first NORM regulations were promulgated in the states of Louisiana and Mississippi.

There are only a few regulations published by agencies to guide the handling or disposal of NORM-contaminated materials, except for uranium mill tailings, which are regulated by the *Nuclear Regulatory Commission (NRC)* and *Agreement States*. There are presently 29 Agreement States and 21 Nonagreement states (Fig. 1-1). NORM generally falls below the dose levels regulated by the NRC. States have been promulgating their own regulations, due to the vacuum or gap not addressed by the NRC. The lack of regulations prompted a private association known as the *Conference of Radiation Program Control Directors (CRPCD)* to initiate and propose guidelines for the regulation of NORM. These guidelines are still in a draft stage and, as noted earlier, some states are waiting for their final publication. The states that have promulgated regulations have required industries to be licensed to extract, mine, benefit, process, use, transfer, or dispose of NORM-contaminated equipment or materials, unless exempted by other regulations. The potential risks to human

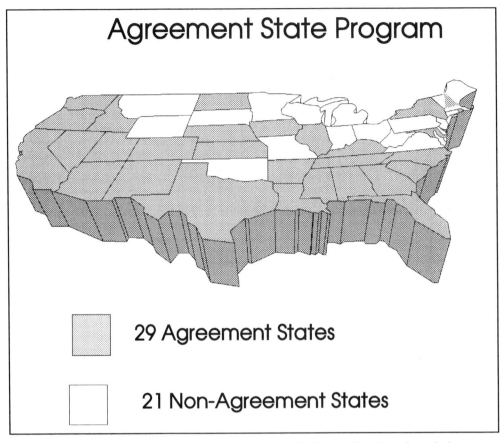

Figure 1-1. Agreement and Nonagreement states governed by the Nuclear Regulatory Commission.

health and safety are largely unidentified. Environmental natural resource damage from NORM has not been quantified to any degree.

The *American Petroleum Institute (API)* and, more recently, the *Gas Research Institute (GRI)* have written position papers on research conducted on the NORM issue.[4,5] API has gathered more than 36,000 NORM data measurements nationwide, and the GRI has recommended a data collection research program. The *New York*

Times reported in 1990 that NORM had been found in every oil- and gas- producing state in pumps, pipes, and storage tanks.[6]

NORM regulations are emerging from these and other discoveries, bringing forth high environmental impacts on a cross section of industries. The regulatory arena is expanding, due to a very well-endowed and very widespread political constituency. It is expected that NORM will become an increasingly visible environmental issue, since it is perceived as a threat to the health and safety of workers, and indeed is a threat, if not properly handled. Regulation will follow closely on the heels of this increased visibility. Disposal of low-level radioactive waste is another giant and looming issue. NORM has a long half-life, and no community wants it stored in its "backyard." Several schemes have been devised for disposal, which will be discussed in more detail in later chapters. Suffice it to state that NORM is a growing regulatory concern and will more than likely become linked to the federal regulatory chain, involving the EPA and OSHA, through reauthorizations and new categories of regulation. States are leading the way, but the federal regulatory effort will spin out of these early efforts and finally be linked to other environmental laws.

CHAPTER 1
REFERENCES

1.   Gray, P.R., "Regulations for the Control of NORM," *Transactions*, SPE/EPA Environmental Exploration and Production Conference, separate paper, 1993.

2.   Rutherford, G.J. and Richardson, G.E., "Disposal of Naturally Occurring Radioactive Material From Operations on Federal Leases in the Gulf of Mexico," *Transactions*, PR/EPA Environmental Exploration and Production Conference, p. 101-107, 1993.

3.   Smith, A.L., "Radioactive Scale Formation," *Journal of Petroleum Technology*, p. 697-706, June, 1987.

4.   American Petroleum Institute, Bulletin E–2; *Management of Naturally Occurring Radioactive Material (NORM) in Oil and Gas Production*, April, 1992.

5.   Spaite, P.N. and Smithson, G.R., *Final Report; Technical and Regulatory Issues Associated with Naturally Occurring Radioactive Material (NORM) in the Gas and Oil Industry*, Gas Research Institute, April, 1992.

6.   Schneider, K., "Officials Examining Danger of Radiation Found in Oil Fields," *The New York Times*, National, December 3, 1990.

7.   *Nuclear Waste News*, "Mississippi Suits Could Decide Oil Companies' Liabilities for Cleanup of Radioactive Wastes," Ziff Communications Co., January 10, 1991.

Basic Radioactivity Theory and Concepts

Radioactivity

The Greek word *atom* means *cannot be broken apart*. The Greeks, however, missed the concept of radioactivity. Atoms *can* be broken apart, and they break through the natural mechanism of radioactive decay. The spontaneous decay of naturally radioactive material emits three distinct types of radiation. They include *alpha* and *beta* particles and *gamma* electromagnetic radiation. An alpha particle is simply a helium nucleus and a beta particle is an electron. There are actually five types[4] of radioactive decay, three of which involve electrons. They are listed below:

1. **Alpha emission (α):** emission of a He^4 particle from an unstable nucleus.

2. **Beta emission (β or β'):** emission of a high speed electron from an unstable nucleus.

11

3. **Positron emission (β^+)**: emission of a positron from an unstable nucleus. Positron emission is equivalent to the conversion of a proton to a neutron. It is a particle identical to an electron in mass, but contains a positive charge.

4. **Electron capture (EC)**: the decay of an unstable nucleus by capturing or picking up an electron from an inner orbital of an atom.

5. **Gamma emission (γ)**: emission from an excited nucleus of a gamma photon, corresponding to radiation with a wavelength of about 10^{-12} wavelength.

For the purposes of this book on NORM, no reference will be made to the terms positron or electron capture. These may be simply defined as an electron emission.

The atom is a tiny "ball" of energy and matter consisting of a *nucleus* surrounded by orbiting *electrons*. The nucleus is composed of *nucleons*. The positively charged nucleons are *protons* and the neutrally charged nucleons are classified as *neutrons*. A neutron can be thought of as a combined proton and electron. The weight of nucleons are approximately 2,000 times heavier than electrons.

Electrons not only orbit the nucleus, but some remain within the nucleus and are linked directly with the neutrons. These might be referred to as "nucleon" electrons. Other electrons orbit the nucleus as an organized "cloud." When a nucleon electron is ejected from the nucleus *(beta emission)*, its neutron loses its negative charge and becomes a proton. Only the negative charge of the electron had prevented the neutron from becoming a proton. The nucleus is only one-trillionth of the total volume of an atom. Obviously the atom is mostly open space.

There are energy levels for the electron orbits, just as there are energy levels for the nucleus of the atom. When *electrons* fall into a lower orbit around the

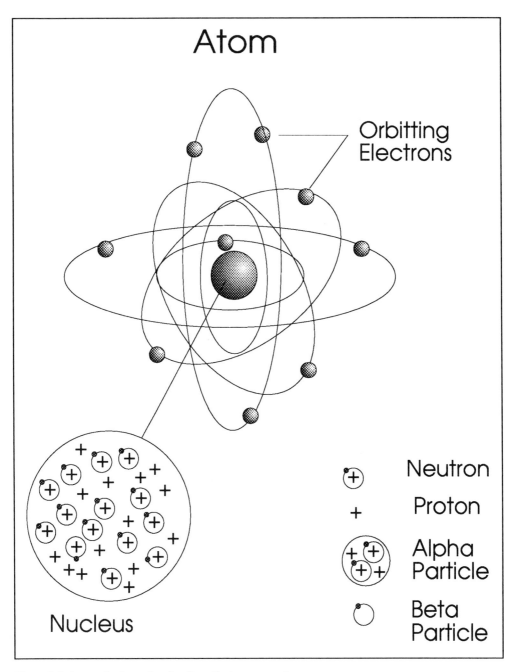

Figure 2-1. A generic atom, showing the salient components.

nucleus they emit *photons of light*. When *nucleons* fall to lower states of energy they emit *photons of gamma radiation*. The nucleus and the orbiting electrons might be thought of as one universe enclosed by another universe. Radiation is emitted mainly from the nucleus of the atom. Other particles are known to exist, but for the purposes of this discussion on NORM, they are not important.

The nucleus of the helium atom is a structure equivalent to the alpha particle. It consists of two protons, two neutrons, and two electrons. It is so large that it will not pass through one sheet of paper;...that is, this sheet of paper and the ink upon it would shield the reader from an alpha particle. Furthermore, because it is a particle, it can be washed from the skin. However, breathing in a material composed of radioactive alpha particles would not be a good idea. A *beta particle* (ejected electron) will penetrate a piece of paper, but will be stopped by a 1/2-in. sheet of aluminum or a thick book. The *gamma ray* can only be stopped by a very dense material, such as a sheet of several inches of lead (see Fig. 2-2). Both alpha and beta particles can be washed from the skin, but gamma rays will easily penetrate the entire body. Alpha and beta particles, however, can enter the body through the respiratory tract, which offers very little resistance to radiation damage. They can also be absorbed through the skin in some circumstances. Skin burns have been noted from beta particles.

Gamma *radiation* travels inversely proportional to the square dependent upon the source activity and mass.

Beta *energy* is dependent upon the isotope's activity, but typically does not exceed a distance of one meter.

Alpha *emissions* are dependent upon the isotope's activity, but they do not travel more than about 4-5 inches.

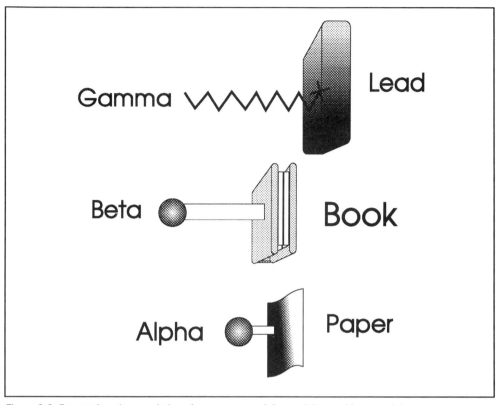

Figure 2-2. Penetrating characteristics of gamma rays, alpha particles, and beta particles.

ISOTOPES

Hydrogen is the simplest of all atoms. It is the element that constitutes the basic building block for water and is the chief component of the stars, the Earth, and the Sun. The hydrogen atom consists of a single proton and a single orbiting electron. The helium atom is the next most complex atom, consisting of two protons and two electrons. Lithium has three protons, and it climbs up the proton ladder of the *Periodic Table of Elements*. Neutral atoms have as many protons in the nucleus as they have orbiting electrons. The number of protons in the nucleus is designated as the *Atomic Number*. Every succeeding element in the Periodic Table of Elements has one more proton than the preceding element (see Periodic Table in Appendix).

This is all very tidy, but the numbers of neutrons in the nucleus do not always follow the rules set forth by the protons and the Periodic Table. Neutrons create variations. Every atom of chlorine, for example, has 17 protons and 17 orbiting electrons, but an *isotope* of chlorine may have more than 17 neutrons. The two major types of chlorine have 35 and 37 times the mass of a single nucleon. In other words, they each have 17 protons, but Cl^{35} has 18 neutrons (17+18 = 35). The Cl^{37} isotope has the required 17 protons and 20 neutrons (17+20 = 37). The atomic mass is also known as the atomic weight; that is, Cl^{37} has an atomic weight of 37. Chlorine is obviously only one of the elements that exist as various isotopes. Even hydrogen, the simplest of all atoms, has three known isotopes. The dual weight hydrogen atom (H^{2}) is designated as *deuterium*, and the triple weight H^{3} is called *tritium*. More than 1,300 isotopes are known to exist within the Periodic Table of Elements.

WHY ARE ATOMS RADIOACTIVE?

The positively charged and closely spaced protons in an atomic nucleus contain very large repulsive electrical charges (see Fig. 2-1). It would appear they would fly away from each other with the speed of light; however, a more formidable force within the nucleus continues to bind the protons together and is known as the *nuclear force*. Both protons and neutrons are bound together by this overpowering nuclear force. The attractive nuclear force is much more complicated than electrical forces and is still not completely understood. The principle portion of the nuclear force is called the *strong interaction*. This force is confined only to protons and neutrons. Protons and neutrons are also known as *hadrons*. The force between the hadrons depends upon their velocity and spin characteristics. If the distance between hadrons doubles, the repulsion of the electrical strength falls by 1/4, and the nuclear attraction falls by only 1/64. As long as protons are in close proximity, the nuclear force can easily overwhelm the electrical repulsion. The larger atomic nucleus begins to lose its nuclear strength on the outer edges of a proton aggregation. The stability decreases as the element nucleus increases in size, weight, and atomic number.

The presence of the neutrons also plays an important role in nuclear stability. Neutrons only attract, not repel, and may be considered "nuclear cement." Most of the first 20 elements in the Periodic Table have equal numbers of protons and neutrons. This equal number is necessary as a counterbalance to instability. More "nuclear cement" is required for the heavier elements, and, as a result, most of the mass in the heavier elements is composed of neutrons. For example, U^{238}, which has 92 protons, possesses 146 neutrons (92+146 = 238). If the uranium nucleus were to have equal numbers of protons and neutrons, it would immediately fly apart, due to the repulsive forces of the positively charged protons. Relative stability can only be achieved by the addition of more neutrons. This stability is tenuous, even with 146 neutrons acting as nuclear cement. The protons begin a steady repulsion around the edges of the nucleus and erosion of alpha and beta particles is inevitable, creating an "atmosphere" of long-term radiation.

It is known that all nuclei having more than 83 protons are radioactively

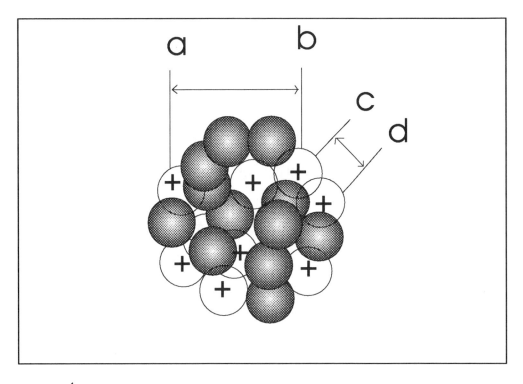

Figure 2-3.[1] The distance between protons and neutrons around the edges of a large nucleus allows the "nuclear cement" to grow weaker between "a" and "b" than between "c" and "d."

unstable. Alpha and beta emission, accompanied by gamma radiation, is part of the shedding process for all elements with more than 83 protons. The actual mechanism for this process involves quantum mechanics and is beyond the scope of the conceptual explanation contained in this book. This process of shedding is known as *radioactivity*. It is dangerous to human health and safety at increasing levels of exposure. All individuals have different tolerance levels to radiation exposure, but some levels are lethal, despite those differences. Figure 2-3 depicts the stability concept of the nucleus in terms of distance.

THE CONCEPT OF "HALF-LIFE"

The radioactive decay, or erosion of an element, is measured in terms of a specific time frame, known as the *half-life*. The half-life of radioactive material is the time for half of the active atoms of any given volume to decay. For example, radium has a half-life of 1,602 years, which means that after 1,602 years, one-half of the original radium will remain. One-half of the remaining radium will be converted into another element over the next 1,602 years, leaving only one-quarter of the original volume of radium atoms after 3,204 years. The other three-quarters of radium has been converted to lead through a series of *disintegrations*. After 20 half-lives, an initial quantity of radioactive atoms will be diminished one-million-fold. The isotopes of some elements have a half-life of less than a second of time, while U^{238} has a half-life of 4.5 billion years. Each radioactive element has its own characteristic half-life. The rates of decay are remarkably constant and are not affected by external physical or chemical conditions. For example, wide temperature and pressure extremes, as well as wide variations in electrical and magnetic fields, have no detectable effect on the rate of decay. Figure 2-4 graphically depicts the concept of half-life.

THE MEASUREMENT OF HALF-LIFE

Half-life is measured in the laboratory and calculated from the rate of decay using radiation detectors. There are four major detectors used to quantify radiation: Geiger-Muller counters, cloud chambers, bubble chambers, and scintillation counters. The field methods employed and calibrations will be discussed in a later chapter.

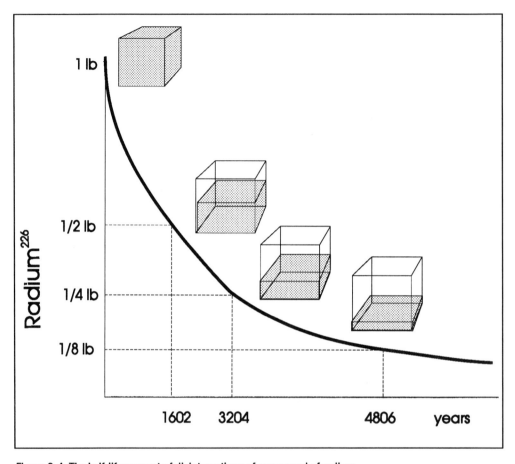

Figure 2-4. The half-life concept of disintegrations of one pound of radium.

The half-life is calculated by the following formula:

$$k = \text{rate} / N_t$$

k =   radioactive decay constant

rate = radioactive decay rate

N_t =   number of radioactive nuclei at time t

EXAMPLE: Ra^{226} has a molar mass of 226 grams and a decay rate of 3.7×10^{10}

disintegrations per second. A mole of a substance contains Avogadro's number (6.02×10^{23}) of molecules or formula units.[4] A one-gram sample of radium[226] contains the following number of nuclei:

$$1.0 \text{ g Ra}^{226} \times \frac{1 \text{ mol Ra}^{226}}{226 \text{ g Ra}^{226}} \times \frac{6.02 \times 10^{23} \text{ Ra}^{226} \text{ nuclei}}{1 \text{ mol Ra}^{226}} =$$

$$= 2.7 \times 10^{21} \text{ Ra}^{226} \text{ nuclei}$$

Substituting this into $k = \text{rate} / N_t$

$$k = \frac{3.7 \times 10^{10} \text{ nuclei/sec}}{2.7 \times 10^{21} \text{ nuclei}} = 1.4 \times 10^{-11}/\text{second for the rate of decay.}$$

The following formula will take the rate of decay, k, and apply it to another formula to calculate the *half-life* ($t_{1/2}$).

$$t_{1/2} = 0.693 / k$$

TRANSMUTATION OF THE ELEMENTS

As alpha and beta particles are emitted from an atomic nucleus, a different element is formed. The alteration of one chemical element to another by radioactive decay is known as *transmutation*. For example, let us examine U^{238} once again. This uranium isotope has 92 protons and 146 neutrons. When an alpha particle is ejected, the nucleus is reduced by two protons and two neutrons (helium nucleus). The remaining 90 protons and 144 neutrons left behind become the nucleus of a new element, thorium (Th^{234}). The reaction is displayed graphically in Figure 2-5.[1]

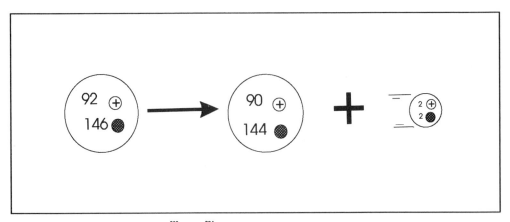

Figure 2-5. The transmutation of U^{238} to Th^{234} through the ejection of an alpha particle.

Note that an arrow is used to show the transmutation of U^{238} into Th^{234}. As this reaction occurs, energy is released in the form of gamma radiation and in the form of kinetic energy (energy of motion) of an alpha particle and, in part, kinetic energy of the Th^{234} atom. Notice that the mass numbers at the top of the elements balance (238 = 234 + 4), and that the atomic numbers at the bottom also balance (92 = 90 + 2). Since this is a nuclear reaction, one can ignore the numbers referring to the orbiting electrons.

The product of the reaction in Figure 2-5 is Th^{234}, which is also radioactive. This new element emits a beta particle when it decays. Recall that a beta particle is a nuclear electron, not an orbital electron. The *nuclear* electron emanates from the nucleus of a neutron. One may envision a neutron as a combined proton and electron. Thus, when the nuclear electron is released, the neutron becomes a proton. A neutron is ordinarily stable when it is locked within the nucleus of an atom; however, a free

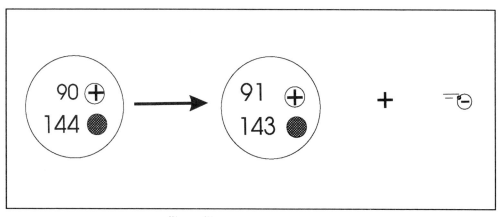

Figure 2-6. The disintegration of Th²³⁴ to Pa²³⁴, plus a beta particle.

neutron is radioactive and has a half-life of 16 minutes. It decays into a proton by beta emission. In the case of Th^{234}, which contains 90 protons, beta emission leaves it one less neutron and one more proton. The newly created nucleus has 91 protons and is no longer thorium, but the element *protactinium*. Although the atomic number has increased by one, the mass remains the same. The nuclear equation is depicted graphically in Figure 2-6.[1]

The previous example (Fig. 2-5), depicts the fact that when an element ejects an *alpha particle* from the nucleus, the *atomic number decreases by 2*. The atom is changed to a lower element in the Periodic Table. On the other hand, when an element ejects a *beta particle* (nuclear electron) its *atomic number increases by 1*. The atom in this later case is changed to a higher element in the Periodic Table. Thus, radioactive elements decay both up and down the Periodic Table. Figure 2-7 depicts the radioactive decay series of U^{238} in alpha and beta decay steps.

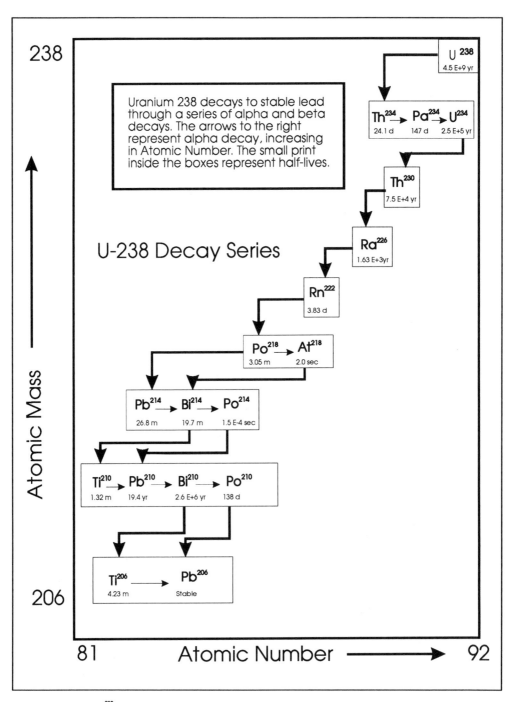

Figure 2-7. Uranium^{238} decay series.

The following table is designed to help you in translating scientific notation to some objects that are familiar by comparing the very large and the very small to one meter.

Table 2-1. The relative sizes of various objects to the meter.[2]

Approximate Length or Distance as Compared to One Meter	Object Description
$1 - 2 \times 10^{26}$	Cosmic horizon
3×10^{25}	Most distant galaxy photographed
$5 - 6 \times 10^{20}$	Diameter of the Milky Way
3×10^{16}	Distance to Alpha Centauri, nearest star to Earth
9.463×10^{15}	One light year
5.9×10^{12}	Distance from the Sun to the edge of the planetary system
1.49×10^{11}	Distance from the Sun to Earth
1.391×10^{9}	Diameter of the Sun
1.2714×10^{7}	Polar diameter of the Earth
4×10^{6}	Distance from New York to Los Angeles
1.8	Homo sapiens sapiens
1.0	**Meter**
7×10^{-3}	Length of a small insect
1×10^{-10}	One angstrom
1×10^{-14}	Diameter of an atom
2×10^{-17}	Diameter of the nucleus of Fe^{56}

Table 2-2 lists all of the elements in the Periodic Table with some of their parameters.

Table 2-2. List of chemical elements.[3]

Atomic Number	Symbol	Name	Atomic Number	Symbol	Name
0	n	neutron	52	Te	tellurium
1	H	hydrogen	53	I	iodine
2	He	helium	54	Xe	xenon
3	Li	lithium	55	Ca	cesium
4	Be	beryllium	56	Ba	barium
5	B	boron	57	La	lanthanum
6	C	carbon	58	Ce	cerium
7	N	nitrogen	59	Pr	praseodymium
8	O	oxygen	60	Nd	neodymium
9	F	fluorine	61	Pm	promethium
10	Ne	neon	62	Sm	samarium
11	Na	sodium	63	Eu	europium
12	Mg	magnesium	64	Gd	gadolinium
13	Al	aluminum	65	Tb	terbium
14	Si	silicon	66	Dy	dysprosium
15	P	phosphorus	67	Ho	holmium
16	S	sulfur	68	Er	erbium
17	Cl	chlorine	69	Tm	thulium
18	Ar	argon	70	Yb	ytterbium
19	K	potassium	71	Lu	lutetium

Atomic Number	Symbol	Name	Atomic Number	Symbol	Name
20	Ca	calcium	72	Hf	hafnium
21	Sc	scandium	73	Ta	tantalum
22	Ti	titanium	74	W	tungsten
23	V	vanadium	75	Re	rhenium
24	Cr	chromium	76	Os	osmium
25	Mn	manganese	77	Ir	iridium
26	Fe	iron	78	Pt	platinum
27	Co	cobalt	79	Au	gold
28	Ni	nickel	80	Hg	mercury
29	Cu	copper	81	Ti	thallium
30	Zn	zinc	82	Pb	lead
31	Ga	gallium	83	Bi	bismuth
32	Ge	germanium	84	Po	polonium
33	As	arsenic	85	At	astatine
34	Se	selenium	86	Rn	radon
35	Br	bromine	87	Fr	francium
36	Kr	krypton	88	Ra	radium
37	Rb	rubidium	89	Ac	actinium
38	Sr	strontium	90	Th	thorium
39	Y	yttrium	91	Pa	protactinium
40	Zr	zirconium	92	U	uranium
41	Nb	niobium	93	Np	neptunium
42	Mo	molybdenum	94	Pu	plutonium
43	Tc	technetium	95	Am	americium
44	Ru	ruthenium	96	Cm	curium
45	Rh	rhodium	97	Bk	berkelium

Atomic Number	Symbol	Name	Atomic Number	Symbol	Name
46	Pd	palladium	98	Cf	californium
47	Ag	silver	99	Es	einsteinium
48	Cd	cadmium	100	Fm	fermium
49	In	indium	101	Md	mendelevium
50	Sn	tin	102	No	nobelium
51	Sb	antimony	103	Lw	lawrencium

Table 2-3 lists a few of the radioactive isotopes, giving their half-life and decay type.

Table 2-3. Commonly available radionuclides.[3] (D.R.) = Daughter Radiation.

Radionuclide	Half-Life	Radiation Type
Americium^{241}	458 years	α, γ, e^-
Antimony^{122}	67 hours	β^-, β^+, γ
Antimony^{124}	60 days	β^-, γ
Antimony^{125}	2.7 years	β^-, γ, e^-
Argon^{37}	35 days	γ
Arsenic^{74}	17.9 days	β^-, β^+, γ
Arsenic^{76}	26.5 hours	β^-, γ
Arsenic^{77}	38.7 hours	β^-, γ
Barium^{131}	12 days	γ, e^-
Barium^{133}	7.2 years	γ, e^+
Barium^{137}	2.55 minutes	γ, e^-
Barium^{140}	12.8 days	β^-, γ, e^-

Radionuclide	Half-Life	Radiation Type
Beryllium[7]	53 days	γ
Bismuth[207]	30 years	e^-, γ
Bismuth[210]	5.01 days	α, β^-, γ
Bromine[82]	35.34 hours	β^-, γ
Cadmium[109]	453 days	e^-, γ
Cadmium[115]	53.5 hours	β^-, γ
Cadmium[115m]	43 days	β^-, γ
Calcium[45]	165 days	β^-
Calcium[47]	4.53 days	β^-, γ
Carbon[14]	5,730 years	β^-
Cerium[141]	33 days	β^-, e^-, γ
Cerium[144]	284 days	β^-, e^-, γ
Cesium[131]	9.7 days	γ
Cesium[134]	2.05 years	β^-, γ
Cesium[137]	30.0 years	β^-, e^-, γ
Chlorine[36]	3.1×10^6 years	β^-, γ
Chromium[51]	27.8 days	e^-, γ
Cobalt[57]	270 days	e^-, γ
Cobalt[60]	5.26 years	β^-, γ
Copper[64]	12.8 hours	$e^-, \beta^-, \beta^+, \gamma$
Dysprosium[159]	144 days	e^-, γ
Erbium[169]	9.4 days	e^-, β^-, γ
Europium[152]	12 years	$\beta^-, \beta^+, e^-, \gamma$
Europium[154]	16 years	β^-, e^-, γ
Europium[155]	1.81 years	β^-, e^-, γ
Gadolinium[153]	242 days	e^-, γ

Radionuclide	Half-Life	Radiation Type
Gallium[68]	68.3 minutes	β^+, γ
Gallium[72]	14.1 hours	β^-, γ
Germanium[71]	11.4 days	γ
Gold[195]	183 days	e^-, γ
Gold[198]	75.6 hours	β^-, e^-, γ
Hafnium[181]	42.5 days	β^-, e^-, γ
Holmium[166]	26.9 hours	β^-, e^-, γ
Hydrogen[3]	12.3 years	β^-
Indium[113]	100 minutes	e^-, γ
Indium[114]	72 seconds	β^-, β^+, γ
Indium[114m]	50.0 days	e^-, γ (D.R.)
Iodine[125]	60 days	e^-, γ
Iodine[129]	1.7×10^7 years	β^-, e^-, γ
Iodine[130]	12.4 hours	β^-, γ
Iodine[131]	8.05 days	β^-, e^-, γ
Iridium[192]	74.2 days	β^-, e^-, γ
Iridium[194]	17.4 hours	β^+, γ
Iron[55]	2.6 years	γ
Iron[59]	45 days	β^-, γ
Krypton[85]	10.76 years	β^-, γ
Lanthanum[140]	40.22 hours	β^-, γ
Lead[210]	21 years	$\alpha, \beta^-, e^-, \gamma$
Lutetium[177]	6.7 days	β^-, e^-, γ
Magnesium[28]	21 hours	β^-, e^-, γ
Manganese[54]	303 days	e^-, γ
Mercury[197]	65 hours	e^-, γ

Radionuclide	Half-Life	Radiation Type
Mercury^{197m}	24 hours	e^-, γ
Mercury^{203}	46.9 days	β^-, e^-, γ
Molybdenum^{99}	67 hours	β^-, γ
Neodymium^{147}	11.1 days	β^-, e^-, γ
Nickel^{63}	92 years	β^-
Niobium^{95}	35 days	β^-, γ
Osmium^{191}	15 days	β^-, e^-, γ
Palladium^{103}	17 days	γ
Palladium^{109}	13.47 hours	β^-, e^-, γ
Phosphorous^{32}	14.3 days	β^-
Polonium^{210}	138.4 days	α, γ
Potassium^{42}	12.4 hours	β^-, γ
Praseodymium^{142}	19.2 hours	β^-, γ
Praseodymium^{143}	13.6 days	β^-
Praseodymium^{144}	17.3 minutes	β^-, γ
Promethium^{147}	2.62 years	β^-
Protactinium^{233}	27.0 days	β^-, e^-, γ
Protactinium^{234}	6.75 hours	β^-, e^-, γ
Radium^{226}	1,602 years	α, e^-, γ (D.R.)
Rhenium^{186}	90 hours	β^-, e^-, γ
Rhodium^{106}	30 seconds	β^-, γ
Rubidium^{86}	18.66 days	β^-, γ
Ruthenium^{97}	2.9 days	e^-, γ
Ruthenium^{103}	39.6 days	β^-, γ
Ruthenium^{106}	367 days	$\beta^-,$ (D.R.)
Samarium^{151}	87 years	β^-, e^-, γ

Radionuclide	Half-Life	Radiation Type
Samarium^{153}	47 hours	β^-, e^-, γ
Scandium^{46}	83.9 days	β^-, γ
Selenium^{75}	120.4 days	e^-, γ
Silver^{110}	24.4 seconds	β^-, γ
Silver^{110m}	253 days	β^-, e^-, γ
Silver^{111}	7.5 days	β^-, γ
Sodium^{22}	2.60 years	β^+, γ
Sodium^{24}	15.0 hours	β^-, γ
Strontium^{85}	64 days	e^-, γ
Strontium^{87m}	2.83 hours	e^-, γ
Strontium^{89}	52 days	β^-, γ
Strontium^{90}	28.1 years	$\beta^-, \text{(D.R.)}$
Sulfur^{35}	88 days	β^-
Tantalum^{182}	115 days	β^-, e^-, γ
Technetium^{99}	2.12×10^5 years	β^-
Technetium^{99m}	6.0 hours	e^-, γ
Tellurium^{132}	78 hours	β^-, e^-, γ
Terbium^{160}	72.1 days	β^-, e^-, γ
Thallium^{204}	3.8 years	β^-, γ
Thulium^{170}	130 days	β^-, e^-, γ
Tin^{113}	115 days	γ
Tin^{119m}	250 days	e^-, γ
Titanium^{44}	48 hours	$e^-, \gamma, \quad \text{(D.R.)}$
Tungsten^{185}	75 days	β^-
Tungsten^{187}	23.9 hours	β^-, e^-, γ
Uranium^{238}	4.5×10^9 years	$\alpha, e^-, \gamma \quad \text{(D.R.)}$

Radionuclide	Half-Life	Radiation Type
Xenon[133]	5.27 days	β^-, e^-, γ
Ytterbium[169]	32 days	e^-, γ
Yttrium[90]	64 hours	β^-, γ
Yttrium[91]	58.8 days	β^-, γ
Zinc[65]	245 days	β^-, e^-, γ
Zinc[69]	57 minutes	β^-
Zirconium[95]	65 days	β^-, γ (D.R.)

Table 2-4 is a list of some common radionuclides and their specific activities.

Table 2-4. Specific activities of selected radionuclides.[3]

Radionuclide	Half-Life	Specific Activities, Disintegrations per Second	Curies per gram
Hydrogen[3]	12.3 years	3.57×10^{14}	9.64×10^{3}
Carbon[14]	5,730 years	1.65×10^{11}	4.46
Nitrogen[16]	7.2 seconds	3.62×10^{21}	9.79×10^{10}
Sodium[22]	2.60 years	2.31×10^{14}	6.25×10^{3}
Sodium[24]	15.0 hours	3.22×10^{17}	8.71×10^{6}
Phosphorus[32]	14.3 days	1.05×10^{16}	2.85×10^{5}
Sulfur[35]	88 days	1.55×10^{15}	4.2×10^{4}
Chlorine[36]	3.1×10^{5}	1.19×10^{9}	3.21×10^{-2}
Argon[41]	1.83 hours	1.55×10^{18}	4.18×10^{7}
Potassium[42]	12.4 hours	2.23×10^{16}	6.02×10^{5}
Calcium[45]	165 days	6.51×10^{14}	1.76×10^{4}

Radionuclide	Half-Life	Specific Activities, Disintegrations per Second	Curies per gram
Chromium^{51}	27.8 days	3.41×10^{15}	9.21×10^{4}
Manganese^{54}	303 days	2.95×10^{14}	7.98×10^{3}
Iron^{55}	2.6 years	9.25×10^{13}	2.5×10^{3}
Manganese^{56}	2.576 hours	8.03×10^{17}	2.17×10^{7}
Cobalt^{57}	270 days	3.14×10^{14}	8.48×10^{3}
Iron^{59}	45 days	1.82×10^{15}	4.92×10^{4}
Nickel^{59}	8×10^{4} years	2.80×10^{9}	7.58×10^{-2}
Cobalt^{60}	5.26 years	4.18×10^{13}	1.13×10^{3}
Nickel^{63}	92 years	2.28×10^{12}	61.7
Copper^{64}	12.8 hours	1.42×10^{17}	3.83×10^{6}
Zinc^{65}	245 days	3.03×10^{14}	8.2×10^{3}
Gallium^{72}	14.1 hours	1.14×10^{17}	3.09×10^{6}
Arsenic^{76}	26.5 hours	5.77×10^{16}	1.56×10^{6}
Bromine^{82}	35.34 hours	3.99×10^{16}	1.08×10^{6}
Rubidium^{86}	18.66 days	3.01×10^{15}	8.14×10^{4}
Strontium^{89}	52 days	1.04×10^{15}	2.82×10^{4}
Yttrium^{90}	64 hours	2.01×10^{16}	5.44×10^{5}
Yttrium^{91}	58.8 days	9.03×10^{14}	2.44×10^{4}
Molybdenum^{99}	67 hours	1.75×10^{16}	4.72×10^{5}
Technetium^{99m}	6.0 hours	1.75×10^{17}	4.72×10^{6}
Ruthenium^{106}	367 days	1.24×10^{14}	3.36×10^{3}
Iodine^{125}	60 days	6.44×10^{14}	1.74×10^{4}
Iodine^{130}	12.4 hours	1.94×10^{16}	1.94×10^{6}
Iodine^{131}	8.05 days	1.24×10^{16}	1.24×10^{5}
Barium^{133}	7.2 years	1.38×10^{13}	374

Radionuclide	Half-Life	Specific Activities, Disintegrations per Second	Curies per gram
Cesium^{134}	2.05 years	1.30×10^{13}	1.30×10^{3}
Cesium^{137}	30.0 years	3.22×10^{12}	87.0
Barium^{140}	12.8 days	2.70×10^{15}	7.29×10^{4}
Lanthanum^{140}	40.22 hours	2.06×10^{16}	5.57×10^{5}
Cerium^{141}	33 days	1.04×10^{15}	2.81×10^{4}
Cerium^{144}	284 days	1.18×10^{14}	3.19×10^{3}
Praseodymium^{144}	17.3 minutes	2.79×10^{18}	7.55×10^{7}
Promethium^{147}	2.62 years	3.44×10^{13}	929
Tantalum^{182}	115 days	2.31×10^{14}	6.24×10^{3}
Tungsten^{185}	75 days	3.48×10^{14}	9.41×10^{3}
Iridium^{192}	74.2 days	3.39×10^{14}	9.17×10^{3}
Gold^{198}	64.8 hours	9.03×10^{15}	2.44×10^{5}
Gold^{199}	75.6 hours	1.04×10^{16}	2.08×10^{5}
Mercury^{203}	46.9 days	5.07×10^{14}	1.37×10^{4}
Thallium^{204}	3.8 years	1.39×10^{13}	462
Polonium^{210}	138.4 days	1.66×10^{14}	4.49×10^{3}
Polonium^{212}	304 nano-seconds	6.48×10^{27}	1.75×10^{17}
Radium^{226}	1602 years	3.66×10^{10}	0.988
Thorium^{232}	1.41×10^{10} years	4.03×10^{2}	1.09×10^{-7}
Uranium^{233}	1.62×10^{5} years	3.50×10^{8}	9.48×10^{-3}
Thorium^{234}	24.1 days	8.58×10^{14}	2.32×10^{4}
Uranium^{234}	7.1×10^{8} years	7.92×10^{4}	2.14×10^{-6}
Uranium^{238}	4.51×10^{9} years	1.23×10^{4}	3.33×10^{-7}
Plutonium^{239}	2.44×10^{4}	2.27×10^{8}	6.13×10^{-3}

The activity of a radioactive source is the number of nuclear disintegrations per unit time occurring in the radioactive material. Thus, a sample having an activity of 1.0×10^{-2} Ci is decaying at the rate of $(1.0 \times 10^{-2}) \times (3.7 \times 10^{10}) = 3.7 \times 10^{8}$ nuclei per second. You will recall that a curie (Ci) is 3.700×10^{10} disintegrations per second by definition.

CHAPTER 2
REFERENCES

1. Hewitt, Paul, *Conceptual Physics*, Little Brown and Company, Inc., second edition, 1974.

2. Klein, H.A., *The World of Measurement*, Simon and Schuster, 1974.

3. Lederer, C.M., Hollander, J.M., Perlman, I., *The Table of Isotopes*, sixth edition, John Wiley & Sons, Inc., 1967.

4. Ebbing, D.D., and Wrighton, M.S., *General Chemistry*, third edition, Houghton Mifflin Co., 1990.

CONSTITUENTS OF NORM

THE NATURAL OCCURRENCE OF NORM

Radioactive material is either concentrated and manufactured or is the natural result of planetary formation. The natural formation of radionuclides began within the formation of a star, such as our Sun and its planetary system. This system of rotating planets formed from a mass of rotating dust and gas that swirled into gravitational existence about 4.5 billion years ago. The mass of dust and gas was formed by a previously existing sun (star) or a series of matter left over from several previously existing stars.

Naturally occurring radionuclides are present in the Earth's sedimentary, igneous, and metamorphic crust, along with associated fluids and gases. Twenty-two primordial nuclides have been identified.[1] The longer-lived constituents are depicted in Table 3-1[1] below.

Table 3-1. Some longer lived primordial radionuclides.

Radionuclide	Half-Life (years)	Abundance of Isotopes from Each Element (%)
K^{40}	1.26×10^{10}	0.012
Rb^{87}	4.7×10^{10}	27.83
Th^{232}	1.4×10^{10}	100.00
U^{235}	7.1×10^{8}	0.72
U^{238}	4.5×10^{9}	99.30

There are three important naturally occurring *radioactive series*. They are listed in Table 3-2, below.

Table 3-2. Three important radioactive decay series.

Series Number	Series Designation
1	Uranium Series U^{238}
2	Thorium Series Th^{232}
3	Actinium Series Ac^{235}

Table 3-3. Series of natural decay reactions for U^{238}.

Symbol	Radiation	Half-Life
U^{238}	α	4.51x10^9 years
↓ Th^{234}	β	24.1 days
↓ Pa^{234}	β and I.T.	1.17 minutes
↓ U^{234}	α	2.48x10^5 years
↓ Pa^{234}	β	6.66 hours
↓ Th^{230}	α	7.5x10^4 years
↓ Ra^{226}	α	1.62x10^3 years
↓ Rn^{222}	α	3.82 days
↓ Po^{218}	α and β	3.05 minutes
↓ Pb^{214}	β	26.8 minutes
↓ At^{218}	α	2 seconds
↓ Bi^{214}	β and α	19.7 minutes
↓ Po^{214}	α	1.5x10^{-4} seconds
↓ Ti^{210}	β	1.32 minutes
↓ Pb^{210}	β	19.4 years
↓ Bi^{210}	β and α	2.6x10^6 years
↓ Po^{210}	α	138.4 days
↓ Ti^{206}	β	4.23 minutes
↓ Pb^{206}	None	Stable

Tables 3-3, 3-4, amd 3-5 depict the details of the three series listed above in Table 3-2. These naturally occurring radioactive elements have three common and related characteristics. They are outlined below:

1. The first member of each series has a very long half-life in human terms. For example, U^{238} has a half-life of 4.5 billion years (4.5x10^9).

2. Each series has a gas member, which is a different isotope of the element radon.

3. The final product in each series is a stable isotope of lead (Pb).

The uranium series, U^{238}, which comprises 99.9% of the naturally occurring uranium, contains two very important radionuclides. These are radium^{226} (Ra^{226}) and radon^{222} (Rn^{222}). Radium is usually found associated with carbonates and evaporates, such as gypsum and very fine grained limestones (micrites and biomicrites).

Table 3-4. Radioactive decay series for thorium.

Symbol	Radiation	Half-Life
Th^{232} ↓	α	1.39×10^{10} years
Ra^{228} ↓	β	6.7 years
Ac^{228} ↓	β	6.13 hours
Th^{228} ↓	α	1.90 years
Ra^{224} ↓	α	3.64 days
Rn^{220} ↓	α	54.5 seconds
Po^{216} ↓	α	0.16 seconds
Pb^{212} ↓	β and α	10.6 hours
At^{216} ↓	α	3×10^{-4} seconds
Bi^{212} ↓	β and α	60.5 minutes
Po^{212} ↓	α	3×10^{-7} seconds
Ti^{208} ↓	β	3.1 minutes
Pb^{208}	None	Stable

Table 3-5. Series of radioactive decay for actinium.

Symbol	Radiation	Half-Life
U^{225} ↓	α	7.07×10^{8} years
Th^{231} ↓	β	25.6 hours
Pa^{231} ↓	α	3.25×10^{4} years
Ac^{227} ↓	β and α	21.7 years
Th^{227} ↓	α	18.2 days
Fr^{223} ↓	β	21 minutes
Ra^{223} ↓	α	11.7 days
Rn^{219} ↓	α	3.92 seconds
Po^{215} ↓	α and β	1.83×10^{-3} seconds
Pb^{211} ↓	β	36.1 minutes
At^{215} ↓	α	$\sim 10^{-4}$ seconds
Bi^{211} ↓	β and α	2.16 minutes
Po^{211} ↓	α	0.52 seconds
Ti^{207} ↓	β	4.76 minutes
Pb^{207}	None	Stable

Commercial products that are derived from these rocks include plaster, cement, wallboard, and other associated products. An average of Ra^{226} concentration in soil is about 1.0 pCi/gram (pico curies per gram). The radioactive units will be more fully described in a later chapter.

Compounds of Ra^{226} are generally soluble in water, and under some ground water conditions, radium can be produced at the surface through wells and springs. Radium nuclides may remain dissolved at dilute levels, or may precipitate out due to chemical pressure and temperature changes. Radium can occur in either fresh or brine aquifers. Often, oil and gas are produced with associated brine, which may bring radium to the surface. Ra^{226} has a half-life of 1,602 years, as mentioned before. Waste stream disposal of this radioactive material should consider half-life as a long term health and human safety factor. Geologic faults, ground water migration, and well plugging procedures are all affected by the terms of radioactive half-life. It becomes the radioactive "legacy factor" that most concerns our environmental stewardship of this planet. Concentrations of radioactive material buried today, either on the land, in the oceans, or injected into wells, may come back to haunt future generations, if it is not planned properly. That is assuming humanity decides not to blow itself up, or pollute itself into a very low state of civilized existence in the next few hundred years.

$Radium^{226}$ produces an external radiation field, but more importantly, has a gaseous daughter element, $radon^{222}$, which can build up in confined spaces. High radon concentrations exist in fresh water aquifers in many areas of the United States. It is estimated that about 25% of the potable water in the U.S. contains Rn^{222} concentrations of greater 2,000 pCi/liter. It is estimated that approximately 5% of the nation's water supply contains concentrations above 10,000 pCi/liter of Rn^{222}.

Radon gas can be released from water via agitation from pumping and/or heating. There are innumerable ways to release radon to the detriment of human health and safety. Water containing as little as 1,000 pCi/liter of Rn^{222} can produce a vapor phase concentration of 1 pCi/liter in confined spaces. The EPA recommends that confined spaces average no more than 4 pCi/liter on an annual basis.

Radon^{222} is also present in some natural gas and some oils. It is produced and brought to the surface via tubing and product lines, which can become contaminated. Household use of natural gas can also become a safety hazard if sufficient quantities are present. Radiation could become a very widespread hazard to health and human safety in coal-burning electrical generating stations. The long term or latent effects of exposure to low levels of radiation are largely unknown. One can visualize a smokestack with billowing black clouds, but one cannot "see" radiation...out of sight, out of mind, but unfortunately not out of the realm of being a potential health hazard.

Radon gas can be a substantial factor within homes built in certain parts of the United States (see Chapter 4). There is an increasing awareness that radon can be a toxic element found in ground water. It is a colorless and odorless gas which has not been studied to any great degree because of the hazardous risks associated with formulated compounds. The Newark Series of Triassic rocks extends from Connecticut to Georgia and is present at the surface. This time-rock unit contains sedimentary sections that have radioactive elements in excess of normal background levels. The South Carolina Department of Environmental Control in 1992 sampled the water from 192 water wells for radon and found that 66% of the wells tested higher than would be allowed under proposed EPA regulations. The standard is presently referenced at 300 pCi/liter. Housing built upon this series of formations often have problems with radon contamination in confined spaces and ground water. Certain formations in Texas and other western states also have the potential for radon contamination. This will be a growing environmental issue as more information is compiled from radiation surveys. It is fortunate that radon can be purged from confined spaces. It is unfortunate, however, that very few people are aware of the radon problem and where it might exist.

Radon decays into solid isotopes. Two of these isotopes, Po^{214} and Po^{218} (Polonium) , when deposited in the lungs, can severely damage cell tissue and create the catalyst for cancer.

Potassium (K^{40}) is another important, potentially hazardous isotope found in nature. K^{40} can be found in plants, animals, and in human bones. It is widely distributed in nature, with volume concentrations ranging from 0.1% to 3.5% in carbonates (limestones). The bones of an average man or woman contain concentrations of approximately 17 mg of K^{40}. The average annual radiation dose received from this amount of K^{40} is approximately 20 mrems to tissue and 25 mrems to bone. We have evolved to live with this amount through time, but fertilizers concentrate this radioactive material in quantities that might prove dangerous in certain situations. The learning curve is still in the embryonic stages, as it applies to the long-term effects upon humans.

CHAPTER 3
REFERENCES

1. Wilson, R., Hendrick, W.G., "NORM: Occurrence, Handling and Regulations," Hazardous Materials Control Resources Institute Conference, New Orleans, 1992.

2. Considine, D.M., *Van Nostrand's Scientific Encyclopedia*, fifth edition, 1976.

GEOLOGY OF NORM

INTRODUCTION

The geology of NORM is a complex subject, but it is a necessary piece of knowledge to predict the occurrence of NORM at the surface and in the subsurface. NORM can occur in fresh water and brine aquifers. It can be found in oil and gas and associated brine production. NORM is found in soils and surface deposits of carbonates (limestones) and certain evaporates. NORM is found in igneous, metamorphic, and sedimentary rocks both at surface and in the subsurface.

The naturally occurring geologic radioactive isotopes[5] are found in almost all geological materials. K^{40} is the most common isotope found in granites, gneisses, slates, shales, sandstones, and most glacial deposits. Most of the remaining natural radiation stems from additional isotopes including U^{238}, U^{235}, Rn^{222}, Ra^{226}, and Th^{232}. The reader is referred to a table in the Appendix that lists the indoor radon percentages for 35 states.

Sea water contains all of the radioactive elements in various concentrations. Uranium accounts for about 3×10^{-3} mg/liter while radium amounts to about 6×10^{11} mg/liter and radon at 6×10^{-16} mg/liter.[6] It is noteworthy that barium concentration in sea water is about 3×10^{-2} mg/liter and sulphur at 8.85×10^{2} mg/liter. It is unknown if $BaSO_4$ (barite) and radium scale found in oil field tubulars emanates from fossil sea water or injected drilling mud. Research is lacking in this issue, and it is generally assumed that $BaSO_4$ is derived from fossil brine. This assumption may be erroneous. It does seem to be a coincidence that the weighting material in drilling mud is $BaSO_4$.

The top ten most abundant elements in the Earth's crust do not include any radioactive elements, but they do comprise the bottom five. This would lead you to believe that radioactive elements have very little impact on human health, but this is not the case. The issue is the concentration of these radioactive elements in the crust of the Earth and its associated fluids. Thorium is the most abundant of the radioactive elements in the Earth's crust, and uranium is the second most abundant radioactive element. Radioactive isotopes of more common elements have not been studied as to their crustal abundance.

URANIUM

All rocks contain small amounts of uranium in quantities that range between 1-3 parts per million (ppm). One million pounds of rock will contain 1-3 lbs. of uranium, as a background element. Soils derived from these rocks will contain about the same amount of uranium. A few rock types contain more than 1-3 ppm. These include light-colored volcanics, granites, dark shales, phosphate sedimentary rocks, and metamorphic rocks derived from all of the above. These rocks contain as much as 100 ppm uranium. The radioactive rocks underlie many areas in the United States, both at the surface and in the subsurface. A few areas exceed the 100 ppm background figure and often are associated with uranium ore concentrations. The key to understanding the presence of uranium is to visualize the Earth in three dimensions. For example, the Texas Gulf Coast has very low uranium background levels at the surface, but can have very high concentrations in the subsurface, which could then become a risk problem in fresh water as well as oil- and gas-associated brine water. Radon, which is the daughter product of uranium decay, concentrates in the soils above the high areas on the surface and the water in the subsurface.

RADIOACTIVE AREAS OF THE UNITED STATES

High concentrations of radioactive soils occur across the United States in many varied places. The occurrences are dependent upon the underlying bedrock that forms the radioactive soils. Figure 4-1[1] depicts the keyed areas to the description below.

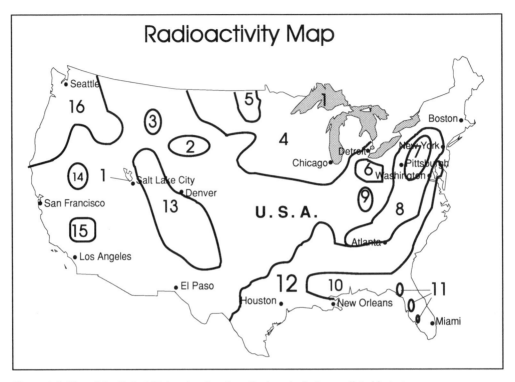

Figure 4-1. Map of the United States showing described geological areas listed in text.

1. **Great Salt Lake:** Water absorbs gamma rays and this becomes a no-data area.

2. **Nebraska Sand Hills:** Wind has separated the lighter quartz sand from the clay and heavier minerals that usually contain uranium. It is a low background area.

3. **The Black Hills:** A core of granites and metamorphic rocks, high in radioactivity, is surrounded by less radioactive rocks. This forms a distinctive pattern.

4. **Pleistocene Glacial Deposits:** This is an area of low surface radioactivity, but is characterized by high radioactivity just below the surface, and carries a high radon potential.

5.   **Deposits of Lake Agassiz:** Clay and silt formed from an ancient, long-evaporated glacial lake have higher radioactive background levels than the surrounding sedimentary material.

6.   **Ohio Shale:** Uranium-bearing black shales occupy this area within a narrow outcrop, which was spread over a large area in west-central Ohio by glaciers.

7.   **Reading Prong:** Uranium-rich metamorphic rocks, associated with numerous small faults, produce elevated radon levels at the surface and in the ground water.

8.   **Appalachian Mountains:** Granites, metamorphics, limestones, and black shales associated with a dense fault and fracture pattern produce higher-than-normal levels of radioactive uranium and radon.

9.   **Chattanooga and New Albany Shales:** These black shales, high in organic matter, contain elevated levels of radioactive elements. These shales extend over portions of Ohio, Kentucky, and Indiana.

10.   **Outer Atlantic and Gulf Coastal Plain:** This area is characterized by unconsolidated sands, silts, and clays and is very low in radioactive elements, with some exceptions. The shallow subsurface contains some elevated levels of radioactive minerals, which affect the ground water.

11.   **Phosphatic Rocks, Florida:** These rocks are high in phosphate and associated uranium. They are mined for the phosphate in a few areas of Florida.

12.   **Inner Gulf Coastal Plain:** This area is characterized by sandy sediments, which contain the radioactive mineral glauconite. Glauconite is nothing more than fossil fecal pellets. You will recall that organic matter is often associated with radioactive minerals and elements.

13.   **Rocky Mountains:** Granites and metamorphic rocks contain higher background

levels of radioactive minerals than the surrounding intermontaine sedimentary basins. This sets up the radon scenario at the surface and in the shallow ground waters.

14. **Basin and Range:** Granitic and volcanic rocks in the ranges alternate with basins filled with alluvium shed from these mountains in a quilt pattern. Both the basins and the ranges have elevated levels of radioactive minerals.

15. **Sierra Nevada:** Granites containing high uranium are particularly noticeable in east-central California.

16. **Northwest Pacific Coastal Mountains and Columbia Plateau:** This is a broad area characterized by low levels of radioactivity, despite the fact it is covered by volcanic basalts.

WATER AND SWAMPS

Surface water and swamps block the effects of surface radioactivity. However, it does not necessarily mean that radioactive elements do not exist within the associated sediments. Soils that are saturated with water also dampen the ability of radioactive measurements. The average amount of radiation in areas of southern Louisiana and parts of Florida may be deceptive at the surface.

THE GENESIS OF RADON

The reader will recall that the daughter products of uranium are radium and radon. Rocks contain grains of minerals, either loosely packed or very tightly packed. The location of the radium atom in the mineral grain (the proximity of the atom to the surface of the grain), and the direction of recoil of the radon atom, determine whether or not the newly created radon atom enters the pore space between the mineral grains, or embeds itself into the surface of the grain. Each atom of radium decays by ejecting from its nucleus an alpha and a beta particle. When the alpha particle is ejected, the newly created radon atom recoils in the opposite direction. Figure 4-2 depicts the reaction.

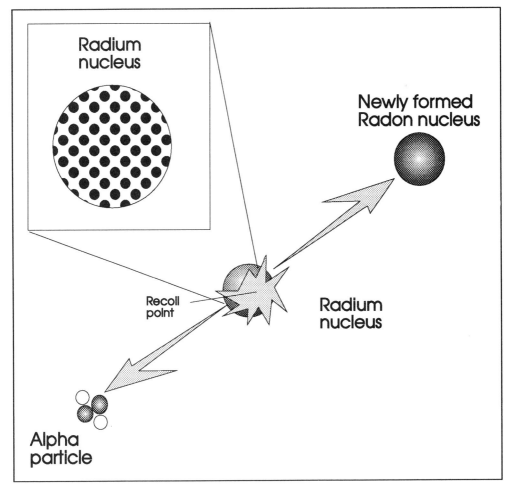

Figure 4-2. A radium atom decays to radon, releasing an alpha and beta particle.

The pore space is generally filled with water. However, near the surface it can just be filled with air in the vadose zone. Only about 10-50% of the radon atoms produced as a decay product end up in the pore spaces of rocks. Most soils contain between 0.33 and 0.1 pCi/gram of radium and 200 to 2,000 pCi/liter of radon in the pore spaces. Figure 4-3 depicts the presence of radon in pore spaces.

The natural geologic sources of radon have recently been studied in some detail. The average American receives annual radiation doses from several sources, both natural and man-made. Radon accounts for about 40% of the average radiation dose received over a one-year period.[5] Gamma radiation from various natural geologic sources contributes 15%. Internal human body K^{40} and U^{238} sources account for another 15%, while cosmic radiation contributes about 12%, and medical dosage, about 17%. Another 1% emanates from unknown and miscellaneous sources.

RADON AND PERMEABILITY

Radon gas can move rather easily through pore space. It has a great deal of mobility when compared to radium, thorium, and uranium, which are fixed within the mineral lattice of the rocks. Radon moves through fractures, small faults, and pore spaces to the surface and can be concentrated in confined spaces, such as a basement or a room within a house. Due to its mobility, radon can travel a considerable distance before it decays. The amount of water within the pore space controls and subdues some of the movement of radon. The permeability of the soil is a very important factor. Highly permeable soil allows mobility, while less permeable soils,

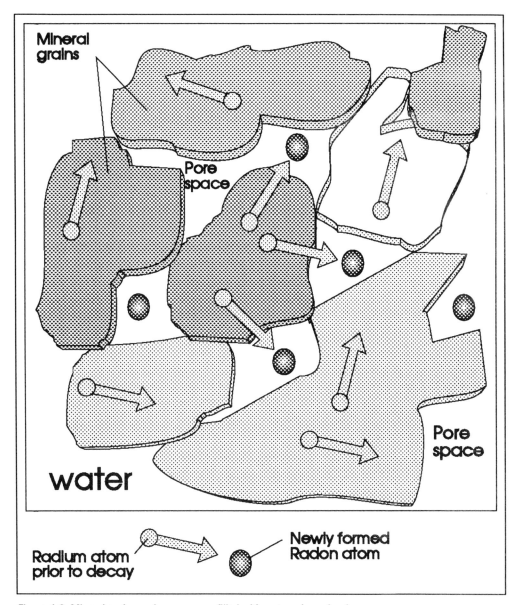

Figure 4-3. Mineral grains and pore spaces filled with water, air, and radon.

such as clays, will impede mobility. Radon in water moves much slower than in air. The distance that radon can move before most of it decays is less than an inch in water-saturated soils, and as much as 6 ft. through dry rocks and soils. The homes that have been built in high-risk radon areas with high and dry permeabilities pose the highest threat to the latent effects of radiation exposure. If the radon levels are relatively low (200-2,000 pCi/liter), the permeability of these areas permits radon-bearing air to move greater distances and concentrate within the confines of a house or building. Figure 4-4 depicts the relationship of radon to permeability.

Radon can move into a building through several pathways, including loose-fitting pipes, sumps, floor drains, minute fractures in concrete, and all other openings to the soil profile. Radon can also enter the confines of a building through water systems. Public water supplies that rely upon surface water do not have the problem that water wells may pose. In the Texas and Louisiana Gulf Coast, ground water is very plentiful to depths that exceed 2,000 ft. These deep wells often set a *screen* over the freshwater aquifer, which is often hundreds of feet thick. If the production screen is set across a radioactive deposit (which is not uncommon), radon and radium will enter the well and end up in the home or building. Gamma logs are sometimes run to discern lithology and often can pick up highly undesirable radioactive zones that should be cased off from the borehole. However, this is not always the case, and a fair number of public supply water wells in the Gulf Coastal area carry higher-than-normal levels of radon and associated radionuclides.

Radon escapes into homes as people take showers, wash dishes and clothes, or otherwise utilize the radon-containing water. In general, house water with 10,000 pCi/liter contributes about 1 pCi/liter to the level of radon in the air. The EPA

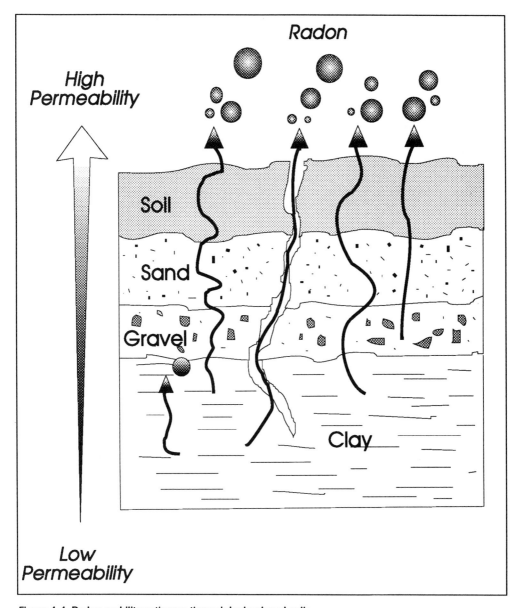

Figure 4-4. Radon mobility pathways through bedrock and soils.

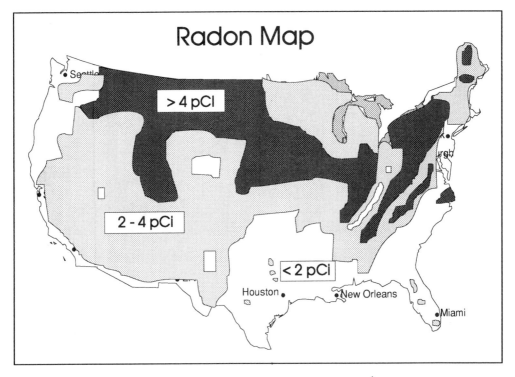

Figure 4-5a-b. Radon map of the U.S. showing three zones of airborne radon.[4]

recommends that no home or building contain more than 4 pCi/liter of radon on an average annual basis. The EPA considers radon to be an extremely dangerous risk for the genesis of cancer.

GEOLOGIC MAPS

Geologic maps, which are available from the United States Geological Survey (U.S.G.S.) and state agencies, can serve as an index to the potential for radon hazards. These maps generally describe the surface rocks, and you can judge from the previous discussions in this book whether or not the surface geology would be a contributing factor to the presence of radon. The geology of the subsurface is a little more complicated and the consulting services of a geologist should be utilized when dealing with ground water and wells.

GEOLOGIC REPOSITORIES OF RADIOACTIVE WASTE

The Nuclear Waste Policy Act (NWPA) of 1982 authorizes the U.S. Department of Energy to construct permanent geologic repositories for spent radioactive fuel and weaponry waste.[2] This is high-level radioactive waste. Low-level wastes are in a state of flux as to their disposal. Only one facility at the time of this writing will take NORM waste. It is located in Utah. Moving a drum of NORM from Louisiana to Utah could run as high as $1,800 per drum.

The question arises, what rock-type could isolate radioactive waste for centuries, and how deep should the waste be buried? Future faulting and movement of the Earth through time are also very important factors to consider. Is it possible to create ground water flow regime models for 10,000 to 50,000 years into the future? These questions and many others are now being addressed by geologists. There are several likely candidates for repositories. Salt deposits offer the best hope for encasement. Salt deposits are either bedded between other sedimentary layers or they form free-floating diapirs or domes that reach from depths of 20,000-30,000 ft. to the surface. Bedded salt may be the better choice because domes can continue to grow with time.

Ground water offers the largest threat to migration. At Nevada's Yucca Mountain site, the water table is 1,900 ft. below the bottom of a proposed repository in the Basin and Range Province. Ten thousand years from now that water table could be at the surface, taking with it contaminated high-level radioactive waste.

No community seems to want a depository in its backyard, but with careful geological investigation, areas and sites will be found in the subsurface that will encapsulate radioactive waste for very long periods of time. It is either that, or send it to the moon.

Some of the geologic site requirements are listed below for a safe repository:[2]

1. Confidence that radioactive ground water will take millennia to reach the biosphere.

2. Thorough understanding of the geochemistry of rocks, ground water, and barrier materials at the site.

3. Packaging that will prevent waste dissolving in ground water, at least through its thermally "hot" period (about 300 years).

4. Shafts that can be permanently sealed against leakage.

5. Confidence that rock media through which any radioactive leachate may migrate will absorb the waste.

6. Knowledge of how rocks and ground water at the site will react to stresses from mining and heat from radioactive decay. Will heat force contaminated water up to the surface?

7. Knowledge of the site's climate. Could it change from arid to moist, elevating the water table?

8. Knowledge of rock stability. Are they prone to being folded or faulted, encouraging ground water flow?

9. Confidence that geologic uplift or subsidence is unlikely.

Figure 4-6. Potential subsurface sites for radioactive nuclear waste.

RADIONUCLIDES IN GROUND WATER

The mobility of radionuclides in ground water is primarily controlled by the chemical and physical processes that occur within the aquifer rock-type.[3] Dissolved uranium is more likely to be found in greater quantities in oxidized ground waters, hosted by granitic rocks, or in some cases, high-grade metamorphic rocks. Uranium accumulations occur in sedimentary rocks at the transition between oxidizing and reducing conditions, along a ground water flow path. Thorium is strongly hydrolyzed and nearly insoluble in all waters. Radium may be quite soluble in ground water, particularly in the absence of sulfates. However, radium is also readily absorbed onto mineral surfaces, which can impede its mobility in ground water. Radon is very soluble in ground water and is not easily absorbed onto mineral grains.

Granitic rocks usually have the highest associated ground water radon levels. Ground water associated with low-grade metamorphic and sedimentary rocks have lower values of dissolved radionuclides, except at the oxidizing-reduction interface.

CHAPTER 4
REFERENCES

1.   Otton, J. *The Geology of Radon,* U.S.G.S., 1992.

2.   AIPG, *Radioactive Waste,* American Institute of professional Geologists, 1985.

3.   Gundersen, L.C.S., Wanty, B.W., "Field Studies in Rocks, Soils and Water," U.S.G.S., Bulletin 1971, 1991.

4.   Duval, J.S., Jones, W.J., Riggle, F.R., and Pitkin, J.A., "Equivalent Uranium Map of the Conterminous United States," U.S.G.S., Open File Report 89-478, 1989.

5.   American Institute of Professional Geologists, *The Citizens' Guide to Geologic Hazards,* 1993.

6.   Lide, D., *Handbook of Chemistry and Physics,* 73rd Edition, CRC Press, 1992.

CHAPTER 5

NORM RADIATION UNITS

UNITS OF MEASUREMENT

This chapter will introduce the confusing array of units of measurements associated with radioactive material. It will also serve as a reference to the material that will follow, as well as a few units found in previous chapters of the book. The first, and perhaps the most fundamental definition, is found in Table 5-1 outlined below. This table depicts the meaning of a few suffixes, which will often be found in scientific literature.

The metric term "gram" is often used with the Table 5-1 prefixes, such as "milligram" or "kilogram." A gram is about 0.0022 pounds or 0.0352 ounces. Oc-casionally one will come across the term *Angstrom*, which is a measurement that designates a very small object equating to 10^{-10} (10 / 1,000,000,000) meter.

The term *curie* (Ci) is often used as a unit of measurement named after Marie Curie (1867-1934), a native of Poland working in a French laboratory. She discovered and named plutonium and radium.

Two Major Categories of Radiation Units

It would appear that radiation units have been designed by scientists to molli-

Table 5-1. Prefix, abbreviation and associated meaning of measurement.

Abbreviation	Prefix	Multiplier
a	atto-	10^{-18}
f	femto-	10^{-15}
p	pico-	10^{-12} (trillionth)
n	nano-	10^{-9} (billionth)
µ	micro-	10^{-6} (millionth)
m	milli-	10^{-3} (thousandth)
c	centi-	10^{-2} (hundredth)
d	deci-	10^{-1} (tenth)
da	deka-	10^{1}
h	hecto-	10^{2}
k	kilo-	10^{3}
M	mega-	10^{6}
G	giga-	10^{9}
T	tera-	10^{12}
P	peta-	10^{15}
E	exa-	10^{18}

fy, obviate, and generally create a fog of confusion for their fellow scientists and the public at large. This conspiracy was developed to conceal the fact that scientists do not want their colleagues to know that they don't know much about radiation. The reason they don't know much about radiation is that all good scientists watched late night movies on television and discovered that Marie Curie was maimed and died of an overdose of radiation. These young, impressionable scientists thought twice about the hazards of radiation research, so they devised a clever curtain of conspiracy laced with confounding nomenclature. This may be a small exaggeration, but brings forth the following simple and diabolical truth.

Radiation units can be simply classified in two major categories. The first category is based upon nuclear disintegration, counted in units of time of a specific volume of radioactive material. The second major category is based upon the biological effects of radiation on the human body and other life forms. The dual classification system might be defined as:

1. *Nuclear disintegration, per unit time, per volume of radioactive material.*

2. *Biological effects of nuclear disintegration.*

If you will mentally classify radiation as the units are encountered and drop the acronyms into one of the above categories, the fog will begin to evaporate.

The most basic and relatively fixed unit of radiation is the *curie*, defined as the *number of nuclear disintegrations per second*. This number is 3.700×10^{10} disintegrations per second. When it was first defined in 1910, the curie unit did not mention any numbers. In 1953, after a convoluted evolutionary process, the curie was defined as any radioactive source that exhibited 3.7×10^{10} disintegrations per second. Unfortunately, no machine nor any human being has been created to predict and quantify the exact moment and exact number of nuclear disintegrations that will occur in any radioactive element. However, relatively accurate radiation counters were developed, which allowed *Specific Activities* to be quantified for most of the radioactive isotopes, fixed in time and space. In other words, it is a number that can be relied upon as "fixed."

The units that define the biological effects of radiation stem from a much more slippery slope, which "swim" in a sea of statistical distribution curves. They are not set in stone with very narrow limits, as is the case in the nuclear disintegration category. This set of numbers and names stemmed from the history of the atomic bomb in the late 1940s and early 1950s. The *rad, rem, RBE, gray, sievert* and *Quality Factor*, which will be discussed later, are all derived from the biological effects on humans. The *roentgen*, which was devised to measure only X rays, served in a dual role for both a description of ionizing radiation and a dose level for humans. However, since about 1956, the roentgen has been increasingly reserved for the first category, that of a fixed number, even though its definition is somewhat different from nuclear disintegrations. The roentgen will be more fully explored in the next section. The biological units, rad and rem, are related to the curie and the roentgen through a series of "fudge" factors to account for the differing effects that radiation causes upon various portions of the body. If the word "absorbed" is encountered, the reader should suspect the unit is based upon biological effects and factors. These factors are based on statistical research that attempts to quantify a very elusive subject. First, humans are not all the same and some have more radio-resistance than others. Second, the research is not based upon a broad population over a number of years. Conversion of numbers between the above described categories is tenuous and slippery at best.

One last point of confusion must be clarified before we move ahead into the oncoming cloud. There is a new movement toward "International Units" (Standard International Units, SI) versus "other" units. The other units were mainly created by American scientists. This book freely jumps from one system of units to the other to make certain the reader is awake.

The anchor of radiation unit nomenclature must ultimately return to the basics of what happens within the atom during nuclear disintegration. All of the numbers and nomenclature have evolved through time, as the tools of measurement continued to improve. It is an evolution similar to something astronomers call "Hubble's Constant," which has changed several times as telescopes improved. Hubble's Constant kept shifting through time. Hopefully, one day radiation will be defined from a common platform, based upon the physical and natural disintegration within a radioactive atom. Until that day arrives, read a book that walks through the history of radiation nomenclature, entitled *The World of Measurements* by H. A. Klein. This book is referenced at the end of this chapter.

OSHA apparently had the same problem as the rest of the population when called upon to define occupational limits of radiation in 49 CFR 1910.96. Their definition of a rem (roentgens equivalent in man) is quoted below:

"Each of the following is considered to be equivalent to a dose of 1 rem:

(i)   A dose of 1 roentgen due to X rays- or gamma radiation;

(ii)   A dose of 1 rad due to X rays-, gamma, or beta radiation;

(iii)   A dose of 0.1 rad due to neutrons or high energy protons;

(iv)   A dose of 0.05 rad due to particles heavier than protons and with sufficient energy to reach the lens of the eye.

(v)   If it is more convenient to measure the neutron flux ...", etc,   etc.

It is obvious that an OSHA rem is a many splendored thing.

You are bombarded with unexplained and undefined radiation units in all types of literature. Hopefully this chapter will begin to define the scope of the nomenclature problem, if nothing else. I wish you all the luck in the world.

REM, RAD, AND ROENTGEN

Units of radiation can be very confusing when trying to convert from one unit to another. The term rem is often used to describe the dosage that can be imposed upon humans. *Rem* is an acronym for *"roentgens equivalent in man."*[1] It was first defined to describe the effects of atomic weaponry on mankind in the early 1950s. Another fundamental unit is roentgen. The term was named after a radiation pioneer, Wilhelm Konrad Roentgen, who discovered X rays. The roentgen is equal to 1.61×10^{12} ion pairs per gram. The energy required to pull the electrons away from neutrons and keep the resulting ion pairs apart is 8.69×10^{-6} joules per gram in dry air. The roentgen was defined to deal with X rays. It was later used as a standard radioactive measurement until the atomic bomb was unleashed and new units were devised to deal with humans and human dosage. The older radiation detectors were calibrated in roentgens per hour, which is about 3.7 disintegrations per second or 10^{8} disintegrations per minute. The term *dpm* is now used to describe the newly proposed Texas NORM regulations, which of course is *disintegrations per minute.* If this all sounds confusing, it is, but don't get discouraged.

One *roentgen per minute (rpm)* is equal to about 1 Curie or 3.7×10^{10} disintegrations per second. One radiation absorbed dose *(rad)* is roughly equivalent to 1 roentgen. The new Texas rules address *rads* as a limiting factor to radiation. *One rad is also equal to the radiations that deposit 1×10^{-2} joules of energy per kilogram of tissue.*

A rem is obtained by measuring a rad and multiplying by the *Relative Biological Factor (RBE),* which is a cousin to the *Quality Factor.* Some parts of the body are more vulnerable to the effects of radiation than other parts of the body. For example, about 250 rems will cause cataracts to form in the lens of a human eye. An average of 450 rems, exposed to the whole body, will prove to be lethal within a few weeks for the average human being. Most of the new radiation field instruments are calibrated in *micro roentgens per hour (µR/hr) or counts per minute.* The term *rem* is a calculated term, applied to industrial hygiene. However, to be able to read certain tables, graphs, and literature, one must be somewhat familiar with the terminology. *Disintegrations per minute* are roughly equivalent to *counts per minute.* If radiation is detected above *background,* the worker should beware and take precautions, until a laboratory analysis can be ascertained.

Background is measured away from any anomalous radioactive areas and must be measured for each investigation. It generally ranges from about 4 µR/hr to about 12 µR/hr for most areas of the Earth's surface. It may range higher for some areas, and it is a number obtained by considering the geology as well as the wetness of the soil or geologic beds being measured.

QUALITY FACTOR

The *Quality Factor* is the amount of radiation absorbed by the whole body and certain organs, such as gonads, breasts, red bone marrow, lung, thyroid, and bone surfaces in the average person on an annual basis. The Quality Factor for the various parts of the body are given in Table 5-2.

The following equation may help the reader understand this biological concept.

rems = rads x RBE

A *rad (radiation absorbed dose)* is equal to about **1×10^{-2} joules.** A joule is related to heat and work and is defined as the amount of heat necessary to lift the

Table 5-2. The Quality Factor for radiation dosage and the percentage factor to use for a dose conversion to rems for an annual dose.

Organ or Body Part	Quality Factor
Gonads	25%
Breasts	15%
Red Bone Marrow	12%
Lungs	12%
Thyroid	3%
Bone Surfaces	30%
Remainder of the Body	30%

temperature of 1 gram of air-free water from 14.5° C to 15.5° C, when the pressure is 1 standard atmosphere. Beta and gamma radiations have an *RBE* of about 1, while a neutron has an *RBE* of about 5. Alpha radiation has an RBE of approximately 10. The effects of radiation on any one individual will vary with the individual, as will the time of exposure and the type of exposure.

The Quality Factor is best described by an example calculation, which is given below:

EXAMPLE:

Given: Annual Exposure

           Whole Body   1.2 rems

           Gonads   2.0 rems

           Lungs   3.0 rems

           Thyroid   8.0 rems

Dose equivalent $= 1.2(1.0\ [100\%]) + 2.0(0.25\ [25\%]) + 3.0(0.12[12\%]) + 8.0(0.03[3\%]) = 2.3$ rems per year.

BACKGROUND, EXPOSURE LIMITS, AND CONVERSIONS

Listed in Table 5-3 are ranges of background levels of NORM in different settings.

The average occupational dose for workers who deal with radiography is approximately 440 millirem/year (mrem/yr), which is roughly two times the normal dose they might receive otherwise.

The average background radiation in Louisiana ranges from 4.0 µR/hr to about 8.0 µR/hour. Offshore the average background will vary between 0.0 µR/hour to 3.0 µR/hour. Background counts are commonly taken on a helicopter pad, associated with a production platform. This is not very scientific, considering that some pads may be constructed of slightly radioactive steel. NORM levels in pipe scale associated with oil and gas production can range from background levels to as high as 17,500 µR/hour. NORM scale packed in drums can often exceed 13,000 µR/hour.

Exposure limits for an individual from external sources is listed in *Title 10, CFR 20.1* (Code of Federal Regulations for Energy). Occupational exposure for an

Table 5-3. Average background radiation dosage from various sources.

Background Areas	Average Background Levels
Harris County, Texas (Lissie Formation outside)	5-6 µR/hour
Harris County, Texas (Lissie Formation inside)	4- 5 µR/hour
Harris County, Texas (Beaumont Formation outside)	6-10 µR/hour
Onshore Southern Louisiana	5-8 µR/hour
Offshore Gulf of Mexico	0-3 µR/hour

individual is *1.25 rem per quarter* and/or *5 rem per year* for the whole body. Occupational limits are raised to *3.0 rem per quarter* and/or *12 rem per year*, if exposure records are maintained. The total lifetime dose cannot exceed *5 x n x 18*. The term *n* is the age at the *last birthday*. Minors (less than 18) must not be exposed to more than 10% of the occupational limits. The whole-body limits for non-occupational exposure is *500 mrem per year*. The maximum rate in the Louisiana regulations for NORM is not to exceed *25 µR/hour* to trigger a reporting permit.

Table 5-4 lists a few conversions and definitions for radiation.

It should be noted that *gamma* radiation is usually measured in *µR/hour*, while *alpha* and *beta particles* are best measured in *counts per minute*.

Table 5-4. Radiation conversions utilizing various units.

Radiation Units	Equates To These Units
pCi/gram	(µrem/hr) x 2.5
1 becquerel (Bq)	1 disintegration/second
1 roentgen	2.082×10^9 ion pairs/cm^3
1 curie	3.7×10^{10} disintegrations/second
1 curie (Ci)	3.7×10^{10} Bq (disintegrations/ second)
1 pico-curie	0.037 Bq
1 Bq/m^3	2.7×10^{-2} pCi/liter
1 pico-curie/liter	37 Bq/m^3
Working Level (WL)	7.4×10^3 Bq/m^3
Working Level (WL)	2×10^2 pCi/liter
1 joule	1 newton x 1 meter
1 rem	0.01 sievert
1 rad	0.01 joule/kilogram

Radiation Units	Equates To These Units
1 CPM	1 DPM
Becquerel (Bq) SI unit	1 disintegration/second
Gray SI unit	Unit of absorbed radiation dose equal to 1 joule of absorbed energy per kilogram of matter
Roentgen	The amount of X- or gamma radiation that produces ionization resulting in 1 electrostatic unit of charge in 1 cubic centimeter of dry air at standard conditions.
Sievert SI unit	Unit of absorbed radiation dose times the quality factor of the radiation as compared to gamma radiation. It is equal to gray times a Quality Factor and has been also defined as 100 rems.
Pair production	The conversion of a gamma ray into a pair of particles—an electron and a positron. This is an example of direct conversion of energy into matter and is quantified by Einstein's formula $E = MC^2$.
10 milli-sievert SI unit	1 rem
10 milli-gray SI unit	1 rad
1 rad	0.01 gray
1 rad	1 roentgen

CALCULATING A DOSE RATE

Calculating dose rate from a field instrument is usually given in µR/hour and best explained by an example.

EXAMPLE: A piece of oil and gas tubing has been measured with a gamma detector with a reading of 2,000 µR/hour. The surveyor's hand was in contact with the spot measured for approximately *1 minute*. The surveyor's body was 2 feet away from the survey spot for approximately 5 minutes. The reading at 2 feet was 500 µR/hour. What was the total dose?

Dose rate at survey

spot (contact with pipe):   (2000 µR/h)/(60 min/h) = **33.33 µR/minute**

                               (33.3 µR/min) X (1 minute) = **33.3 µR**

At 2 feet (whole body)

for 5 minutes: (500 µR/h)/(60 min/h) = **8.33 µR/minute**

                               (8.33 µR/min) x (5 minutes) = **41.65 µR**

                               then;

                               33.33 + 41.65 = 74.95 µR   total dose

Hopefully, the reader is not in total disarray and brain dead by now. There are three "rules" to remember on how to decipher how radiation is measured. They are listed here.

1. Counts per minute and micro-roentgens per hour equate to *field measurements*.

2. Pico-curies, Working Levels, and becqverels are *laboratory measurements*.

3. Rems, millirems, RBE, Quality Factors, and rads are all related to the *radiation effects on humans*. They are not pure physical measurements. They relate pure physical measurements to statistical "best guesses," based on a patchwork of research.

One last thought should be conveyed at this juncture. If you look for radiation conversions within books devoted to conversions, they are usually nowhere to be found. The reader should begin to make collections of conversion factors when they are found ... which is a rare event.

Chapter 5
References

1. Klein, H.A., *The World of Measurement*, Simon and Schuster, 1974.

2. McGuire, S., and Peabody, C.A., *Working Safely With Gamma Radiography*, Office of Nuclear Regulatory Research, U.S. Nuclear Regulatory Commission, 1986.

CHAPTER 6

INDUSTRIAL SOURCES OF NORM

INDUSTRIAL SOURCES OF NORM[1]

Naturally Occurring Radioactive Material has been noted as a byproduct of non-nuclear technological processes. NORM is entrained in certain process materials as it moves through a manufacturing system. NORM is also commonly produced with brine associated with oil and gas. It may remain in solution, or it may precipitate within the production and/or manufacturing equipment. It may also become a portion of the ancillary production stream prior to refinement.

Table 6-1.[1] The estimated annual waste inventory and concentration of NORM by the primary industries. *MT = metric ton = 1000 kg = 1.1 short ton. ** An empirically derived conversion factor for the decay series equal to 1.82 µR/hour per pCi/gram.

Material	Annual Waste Inventory *	Average Ra[226] Concentration	Exposure Rate**
Phosphate waste	40.00 million MT	33.0 pCi/g	60 µR/hour
Coal waste	43.00 million MT	3.7 pCi/g	6.7 µR/hour
Petroleum waste	0.63 million MT	155.0 pCi/g	282.1 µR/hour
Mineral processing	1.00 billion MT	35.0 pCi/g	63.7 µR/hour
Water treatment	0.26 million MT	16.0 pCi/g	29.1 µR/hour
Geothermal waste	0.07 million MT	160. pCi/g	291.2 µR/hour

The following discussion is formatted to address the primary industrial sources of NORM and associated wastes. Table 6-1 depicts a summary of industrial NORM waste as it relates to radium (Ra^{226}) concentrations.

Volumes of NORM waste present a very impressive figure of low-level radioactive waste. NORM may concentrate within a portion of the production stream and create risks to human health and safety in certain instances. Short-term exposures may not prove to be harmful, but longer-term dosage may create scenarios that will only affect the exposed worker later in life. There are no hard-and-fast guidelines for low-level exposure rates. Equipment often must be cleaned, and this is one of the most dangerous aspects of NORM concentration. Drinking water contaminated with radon may prove to be a health risk over long-term periods. The field of

NORM is so new that very few, if any, studies have been made concerning the long-term effects. NORM-contaminated waste is often included in reclamation processes, due to the bulk quantities that generally are associated with NORM. This opens the pathway for incorporation of NORM into consumer products, accompanied by radionuclide concentration through the reclamation process. For example, decorative steel patio furniture manufactured in Mexico was recently found to be radioactive. NORM-contaminated building materials may increase as a vapor phase of radon, if associated with original or reclaimed material contaminated with NORM. Leachate from the purification of uncommon metals may contain radionuclides and heavy metals. Improper disposal of NORM waste may create an environmental hazard and threaten health and human safety. The quantification of risk concerning NORM is virtually an unknown subject. Ignorance breeds fear, however, and the public is often very ignorant and very fearful.

NORM IN THE PETROLEUM INDUSTRY

NORM has a prominent place in the petroleum industry and it is becoming widely perceived as a burgeoning issue. NORM wastes are found in several production and transportation streams within this industry. It is found in the form of radium, which is co-precipitated with silicates, sulfates, and/or carbonates, in sludges and scale. NORM is often associated with barium sulfate ($BaSO_4$), a common constituent of drilling mud used as a weighting agent. Barium sulfate scale is a problem in deep, hot gas and condensate wells. It has been known to completely plug a tubing string and the production tree in some deep wells. Barite ($BaSO_4$) is often lost within the formations as the well is drilled, and later returns in the production stream. NORM becomes an exponentially larger problem as brine production increases, due to the decline in production of oil and/or gas in some reservoirs. It is unknown at the present time whether the $BaSO_4$ is a product of the drilling of the well, or is indigenous to the formation or reservoir fluids. No correlations have yet been found between ages of reservoir or paleo-environments of deposition. There must be predictive correlations within the subsurface geochemistry, rocks and fluids, but they are not yet known. $BaSO_4$ has an affinity to adsorb and incorporate radium as it precipitates upon flow within tubing, casing, and production equipment. The average radon emanation fraction for scale and sludge is approximately 0.10-0.20%. The emanation fraction is the percent of radon formed as it diffuses into the air and soil. Emanation is a flux measured in atoms cm^2 seconds. $Radium^{226}$ concentrations in the petroleum industry can be as high as 40,000 pCi/gram. The typical concentration, however, is approximately 155 pCi/gram. It is estimated that the petroleum industry generates about 360,000 cubic meters or 0.63 million metric tons of NORM waste each year. These figures could be much higher, due to a severe lack of data.

Natural gas flows to the surface in a highly turbulent condition at very high rates. Radon gas is often admixed with methane (CH_4) as it moves out of the reservoir into the tubing and production equipment. Since radon (Rn^{222}) has a very short half-life (3.8 days), it leaves a converted train of radioactive lead (Pb^{210}) along its path. Pb^{210} has a half-life of 22 years. Thin layers of Pb^{210} are often found on the inside surfaces of piping and equipment. These accumulations have an average thickness of 0.004 inches. The residue affects the ultimate disposal of gas processing equipment. The economics of that disposal can be considerable.

NORM may also accumulate inside tubing strings from oil production, which is usually accompanied by the production of brine. The scale within tubing can vary in thickness from a millimeter coating up to an inch or more. This radioactive scale will also accumulate within separators, valves, and piping associated with oil and brine production. It becomes a very onerous disposal problem. The petroleum production most affected by the NORM problem lies within the geographic limits of the Gulf Coast and Midwest United States. Very little is known about all of the affected areas and associated reservoirs. It is very probable that many new areas will slowly be added to the list as awareness rises. It is probable that all of the producing states will discover the NORM production problem, as suggested by several authors.

RARE EARTH MINERAL PROCESSING PLANTS

Rare earth minerals are found in sand-sized particles and are classified as heavy minerals, which include monzanite, zircon, illmenite, and tantalite. These mineral compounds are utilized in many consumer products. These products include red phosphorus for color television tubes, glass polishing powder, glass composition in camera lenses, microscopes, and magnets. The rare earth minerals processing industry contains an average Ra^{226} concentration of 35 pCi/gram and generates about 1 billion tons of waste each year. Thorium is also a common adjunct to heavy minerals.

WATER TREATMENT PLANTS

Ground water contains a substantial amount of NORM in certain areas, which coincide with broad areas of oil and gas production. The main radionuclide is Ra^{226}, which is water-soluble and is transported through convective flow. The filtration process employed by public water supplies creates a sludge, which accumulates and concentrates radium and uranium in the pore spaces of charcoal beds and is admixed with spent ion exchange resins. The average Ra^{226} concentration in water treatment plants is estimated to be 16 pCi/gram. The water treatment waste generated each year is estimated to be 0.26 million metric tons. Some of the radon gas is passed through to the consumer in unknown quantities.

PHOSPHOGYPSUM MINING

Phosphate is mined from several areas in the United States, principally in Florida. Small traces of uranium, thorium and the decay product, radium, exist within the mined phosphates. These phosphates are utilized for the production of phosphoric acid and elemental phosphorus. The Ra^{226} concentration in the phosphate rocks can be as high as 60 pCi/gram.

Phosphogypsum is a byproduct material produced during the manufacture of phosphoric acid. When phosphogysum fertilizer is produced, most of the radium is retained within the phosphogysum waste stream. Only small amounts mix with the fertilizer. The EPA estimates that annual phosphogypsum waste equals approximately 40 million metric tons.

The production of elemental phosphorus creates a waste containing NORM, which is classified as slag when produced in high-temperature furnaces. Slag is often used as an aggregate in the building, construction, and road contracting business. Ra^{226} concentrations in slag range as high as 60 pCi/gram. The average concentration is assumed to be about 33 pCi/gram.

COAL AND LIGNITE POWER PLANTS

Coal and lignite contain varying amounts of naturally occurring radionuclides, depending upon their origin. The Gulf Coast lignites are relatively high in radionuclides compared to the harder coals of the Appalachians. Coal ash is produced as the coal and/or lignite is burned to generate electricity. Coal ash consists of fly ash, bottom ash, and boiler slag. Approximately 27% of all ash is utilized as an additive to concrete. The remainder of the ash and slag is disposed of in landfills. Concentrations of Ra^{226} in coal ash have been found to be as high as 20 pCi/gram. The average concentration is about 3.7 pCi/gram. Utility companies generate approximately 43 million metric tons of coal ash waste each year.

GEOTHERMAL ENERGY PRODUCTION WASTE

Geothermal energy production is a relatively small generator of NORM and is presently confined to California. Contaminated wastes result from the treatment of spent brines. The estimated waste volume from geothermal production plants is estimated to be 70,000 metric tons per year. The estimated Ra^{226} concentration is about 160 pCi/year. It is interesting to note that many hot springs resorts have relatively high concentrations of radionuclides. In this case, the cure may not be worth the cancer risk posed by these resorts and natural health spas.

CHAPTER 6
REFERENCES

1. Wilson, R., Hendrick, W.G., "NORM: Occurrence, Handling and Regulations," Hazardous Material Control Resources Institute Conference, New Orleans , 1992.

CHAPTER 7

HEALTH RISKS ASSOCIATED WITH NORM

INTRODUCTION

The sources of NORM associated with specific health risks must be addressed from a platform of related NORM research, rather than sources that skirt the full spectrum and scope of NORM. Very large doses of radiation will cause harmful effects within weeks or days, while the long-term latent effects remain clouded in unknown realms of darkness. These delayed effects are the precursor of cancer and genetic defects, which may prove to be the most dangerous of all radiation problems. Scientists do not understand all of the causes of cancer, but DNA (deoxyribonucleic acid) may be damaged and launch cell division to the point that leads to cancer. Radiation may be one of the paths to that initial damage.

The body can apparently withstand a higher dose of radiation through ingestion than inhalation. Little is known, however, about the threshold levels of damage of either. This may be due in part to the variability of humans in their individual radio-resistance. It is prudent to take precautions when working with NORM, especially when it may create an atmosphere that may enter the lungs.

NORM sources and wastes that might be associated with radiation health risks should be addressed within the arena of food, water, consumer products, and industrial processes. Ra^{226} (radium) can be absorbed in soil and plants, progressing through the food chain as plant material eaten by livestock and man.[1] Drinking water may have a relatively high concentration of radium and radon, as previously discussed. The concentration of radium ranges from about 0.01 to 1.0 pCi/liter in fresh water, to about 100 pCi/liter in brine. Radon (Rn^{222}) may also be present in ground water. Some of the consumer products that may contain varying amounts of NORM include cloisonne jewelry (enameled), dentures, camera lenses, natural gas, peat moss, lawn fertilizers, rock wool insulation, glass, lamp mantles, and some decorative metal furniture. Industrial sources of NORM generate the great bulk of the radioactive material, and much of the waste ends up in a landfill, which is then free to migrate into ground water in concentrated quantities. The transportation of NORM industrial waste creates a convective vapor phase flow in such things as airborne ash, as well as in many other things.

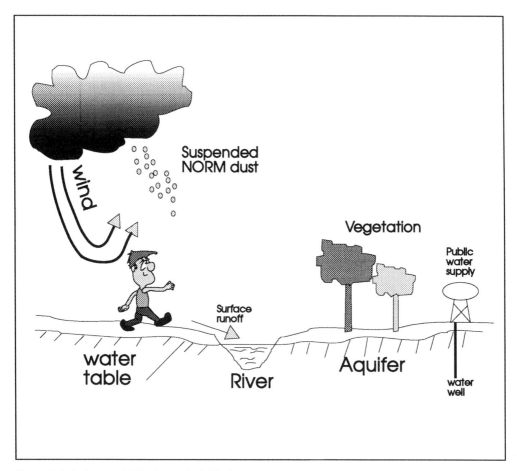

Figure 7-1. Pathways of NORM to an individual worker.

NORM EXPOSURE PATHWAYS

An exposure pathway may be defined as "a single exposure route, like inhalation of dust, ingestion of contaminated food or water, or direct gamma ray exposure."[1] A possible scenario would be "... a combination of pathways based on how a person enters and/or intersects with the facility or waste." Figure 7-1 illustrates the potential pathways, which could simulate a worker at a NORM site.

Figure 7-2 depicts Rn^{222} vapor and ground water exposure pathways. Total exposure to NORM can be viewed as the summation of all pathways, which are shown in Figure 7-2. However, it should be noted that one pathway may be the dominant factor for radiation dose damage.

External exposure to NORM on a single and individual basis, is relatively small over a short time frame. The radiation danger on such a basis stems not from gamma exposure, but internal exposure caused by inhalation of a Ra^{226} or a Rn^{222} particulate matter. Food and liquids can also create relatively immediate radiation damage. Through extensive modeling, these scenarios have been verified as radiation migratory pathways.

Approximately 80-85% of the radium ingested by humans is deposited within bone matter. An average of 40 pCi/gram of Ra^{226} and/or Ra^{228} has been calculated to deliver a dose of 3.5 mrad/year to the bone areas (osteocytes), via alpha and gamma emission. People who worked as radium watch painters were found to have bone cancer if their exposure was higher than 1,000 rads. The coefficient of risk was found to be 6-53 bone sarcomas per million people, per rad average skeletal dose for Ra^{226} and Ra^{228}. In the epithelium of the nasal passages, doses from Ra^{226}, Rn^{220}, and Rn^{222} produces a coefficient of risk for nasal carcinoma development of 64 carcinomas per million persons, per rad from particulate matter.

Biological monitoring of radium watch dial painters in the 1920s revealed that individuals who inhaled and ingested significant quantities of Ra^{226} had increased incidences of lung and bone cancer, compared to the population at large. Moghissi and Carter in 1975[7] calculated that the average vapor phase of radon concentration in dial painting plants was 10 pCi/liter. A model introduced by Harley and

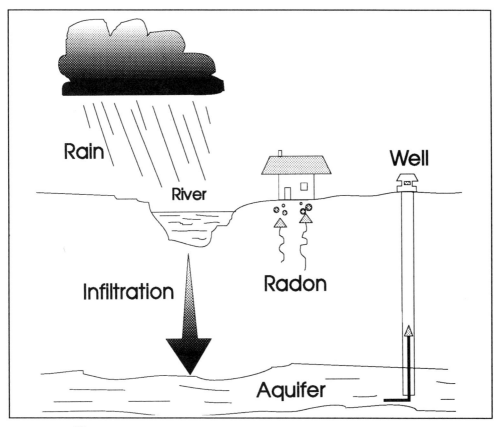

Figure 7-2. Rn^{222} vapor and ground water phase exposure pathways.

Pasternack in 1972 predicted that 168 hours/week of continuous exposure to air-borne radon at 1 pCi/liter would deliver 1,520 mrem/year to the bronchial tubes.[1]

Vapor phase radiation mainly stems from Rn^{222}, which can be quite dangerous. A smaller contribution to airborne radiation is derived from Rn^{220}. Vapor phase concentra-tions of Rn^{220} and Rn^{222} are approximately 1.89 x 10^{-8} pCi/liter and 1.080 pCi/liter, respectively. These decay isotope products of Ra^{226} and Ra^{224} are electrostatically attracted to dust particles with diameters generally less than 0.025 micrometers. During decay, these radionuclides emit alpha, beta, and gamma radiation. Particles of Ra^{226} may also be deposited in lung tissue through inhalation. The dose equivalent from the gas Rn^{222} deposited in lung tissue is approximately 2.0 mrem/year. The segmented portions of the bronchial tubes may receive up to 520 mrem/year. The National Council on Radiation Protection (NCRP) estimates that average annual doses to lung and bronchial epithelium from Rn^{222} and its daughter products are about 2,400 mrem/year.

RADIATION LEVELS ASSOCIATED WITH NORM

An average individual in the United States receives an annual full body dose of approximately 300-500 mrem from natural sources of radiation, which is described as "background". About 20-40 mrem is derived from external gamma radiation through exposure to airborne Rn^{222}. Approximately 130-250 mrem comes from Rn^{222} deposited within the lungs. The remainder of the dosage emanates from various sources, such as ingestion of radiation-bearing food and water, and exposure to building materials, as well as cosmic sources. If a person lives 70 years, the loss of life expectancy from this natural exposure is about 8 days. Table 7-1 depicts the percentage contributions of various radiation sources to an average individual.

MODELS USED TO EVALUATE NORM EXPOSURE RISKS

The primary computer models that are most widely utilized for the evaluation of NORM exposure risks are listed as the following:

Table 7-1. Sources of radiation to humans on an average annual basis.[1]

External Radiation	Elevation	Effective Dose Rate
Cosmic ray dose in U.S.	sea level	27 mrem/yr
	5,000 ft	44-60 mrem/yr
	10,000 ft	85 mrem/yr
Radon in air (breathing).		22-40 mrem/yr
Gamma rays inside a brick and concrete building.	sea level	91-261 mrem/yr
Gamma rays from soils and geological formations.	coastal areas	15-35 mrem/yr
	plateau areas	75-140 mrem/yr
Internal Radiation from Human Tissue		
Radon daughters deposited in lungs.		130-250 mrem/yr
K^{40}		20 mrem/yr
C^{14}		1 mrem/yr
Avg. Total Natural Whole Body Dose Rate		**300-450 mrem/yr**

1. IMPACTS-BRC

2. PATHRAE

IMPACTS-BRC was developed for the calculation of total body exposure. The original FORTRAN code was written by the Nuclear Regulatory Commission (NRC) and may be used with the operating system of a personal computer. A data base is built into the program, which contains information on 85 radionuclides as well as regional rainfall and wind speed information. Simulation of nine separate scenarios are possible. They are listed below:

1. Moving radioactive waste to a disposal site.

2. Digging a foundation on a contaminated site.

3. Living in a ranch style house on a site.

4. Ground water migration of radionuclides.

5. Exposure of workers at the radioactive waste disposal facility.

6. Exposure of radioactive waste at the surface.

7. Incineration of radioactive waste.

8. Recycling of glass and metal.

9. Leachate accumulation and overflow in a landfill.

PATHRAE allows the user to create scenarios through a combination of various factors that affect pathways for radiation to follow to a receptor. The program allows the user to create five on-site and five off-site scenarios. The on-site pathways address ground water contamination and migration to streams and wells. The off-site pathways address ingestion, absorption, and inhalation.

Table 7-2 depicts PATHRAE calculated exposures from direct gamma exposure from dust and radon inhalation for workers at a NORM disposal site. Table 7-3 shows the calculated health effects of NORM exposure by industry on the U.S. population. The two waste sectors with the largest detrimental health effects from radiation are the coal ash and mineral processing industries. The calculations are based upon the excess fatal cancers that result from one year of exposure. The PATHRAE simulations, which did not include radon inhalation, show that workers at NORM storage and/or facilities incurred the highest health risk from direct gamma ray exposure. Dosage rates from dust inhalation were calculated to be about *100-1,000 times less* than gamma ray exposure. The reader may note that this is counter to current wisdom concerning NORM exposure. The simulation may assume the proper use of personal protective equipment, which is not always the case in the real world.

MEASUREMENTS OF HEALTH EFFECTS

It is often found that environmental, health, and safety professionals will refer to a health risk as "10^{-3}," which does not mean much to the average person without some explanation. If *1* is divided by *1000* the result is *0.001* or *1.0 x 10^{-3}*. This might be interpreted as 1 additional fatal cancer case per 1,000 people. These terms or numbers are often found within the discipline of risk assessment, which is becoming a very visible environmental issue and is routinely performed on larger projects. OSHA has been guided by studies that classify occupations as *high risk*, *average risk*, and *low risk*. High risk occupations such as fire fighting or mining have a range of occupational deaths from 20.16 to 27.45 per 1,000 workers. Average risk occupations, including all manufacturing and service employment, ranges from 1.62 to 2.7 deaths per 1,000. Low risk occupations average between 0.07 and 0.48 deaths per 1,000. This translates to 7.0×10^{-5} to 4.8×10^{-4} for low risk occupational deaths. The FDA has interpreted several court decisions to mean that a lifetime risk of 1 in 1 million (1,000,000) is a "deminimus" level of risk, therefore an acceptable risk.[8] This translates into 3 excess cancer cases per year in a population of 227 million people with a life expectancy of 74 years ($10^{-6} \times 227 \times 10^{6}/74$). The EPA has commonly accepted a 10^{-6} health risk for other contaminants. The

Nuclear Regulatory Commission has proposed higher acceptable lifetime risks for persons working with radiation. These have been set at 4×10^{-4} or 4 excess deaths per 10,000 workers. This is compared to the general public which has an acceptable death rate at no more than 5×10^{-3} or 5 persons per 1,000. It is a controversial issue with every contaminant, and it involves "voluntary" versus "involuntary" exposure factors, such as a person who voluntarily works with NORM and those that do not volunteer.

THE HUMAN CELL

The structure of the body is quite complex and it is convenient to deal with health effects at certain levels of organization within the body. The human body consists of many organs, each of which is composed of two or more types of tissue. A tissue is composed of two or more cells. There are four types of tissue in the body, which include the following:

1. Epithelial
2. Connective
3. Muscle
4. Nerve

The job of a tissue is to perform a special body function. The cell is composed of many chemical elements, which include the following:

1. Hydrogen
2. Oxygen
3. Carbon
4. Nitrogen
5. Lesser amounts of other elements

Table 7-2. Worker doses and health effects from storage and/or disposal of NORM utilizing the program PATHRAE. (a) The number of excess fatal cancers (70-year lifetime risk) expected in the total U.S. population as a result of one year of exposure.

Industry	Direct Gamma Exposure (mrem/year) Health Effects(h.e.)	Dust Inhalation Dose (mrem/year) Health Effects (h. e.)	Radon Inhalation Health Effects [a]
Uranium Overburden	6.5×10^1 mrem/yr 2.5×10^{-5} h.e.	2.8×10^{-2} mrem/yr 4.3×10^{-9} h.e.	1.8×10^{-2}
Phosphate Waste	9.9×10^1 mrem/yr 3.8×10^{-5} h.e.	1.2×10^{-2} mrem/yr 1.9×10^{-9} h.e.	1.2×10^{-2}
Phosphate Fertilizer	6.2×10^{-3} mrem/yr 2.4×10^{-9} h.e.	1.7×10^{-5} mrem/yr 2.7×10^{-12} h.e.	unknown
Coal Ash	1.6×10^1 mrem/yr 6.3×10^{-6} h.e.	1.1×10^{-2} mrem/yr 1.6×10^{-9} h.e.	1.4×10^{-4}
Water Treatment Sludge Fertilizer	2.3×10^0 mrem/yr 8.8×10^{-7} h.e.	1.3×10^{-4} mrem/yr 2.0×10^{-11} h.e.	unknown
Water Treatment Sludge Landfill	8.0×10^{-1} mrem/yr 3.1×10^{-7} h.e.	4.8×10^{-5} mrem/yr 7.6×10^{-12} h.e.	1.2×10^{-4}
Mineral Processing Waste	1.5×10^2 mrem/yr 5.9×10^{-5} h.e.	6.9×10^{-2} mrem/yr 1.1×10^{-8} h.e.	2.7×10^{-2}
Oil & Gas Scale-Sludge	6.5×10^2 mrem/yr 2.5×10^{-4} h.e.	4.0×10^{-2} mrem/yr 6.3×10^{-9} h.e.	2.1×10^{-2}
Geothermal Waste	6.7×10^1 mrem/yr 2.6×10^{-5} h.e.	2.5×10^{-2} mrem/yr 3.8×10^{-9} h.e.	9.3×10^{-2}

Table 7-3. Summary of cumulative health effects in the U.S. from one year of exposure in terms of the number of excess fatal cancers (70-year life span) expected in the total U.S. population.

Waste Sector	Number of Sites for a 20-Year Inventory	Health Effects [a]
Uranium Overburden	1.4×10^1	2.4×10^{-1}
Phosphate Waste	1.5×10^1	5.2×10^{-1}
Phosphate Fertilizer	9.4×10^5	9.2×10^{-1}
Coal Ash	1.3×10^3	1.2×10^1
Water Treatment Sludge Fertilizer	4.4×10^2	7.0×10^{-2}
Water Treatment Sludge Landfill	2.3×10^2	1.8×10^{-3}
Mineral Processing Waste	6.7×10^2	1.7×10^0
Oil & Gas Scale and Sludge	1.0×10^1	5.6×10^{-2}
Geothermal Waste	2.0×10^0	1.5×10^{-2}

The components of the cell deserve some attention, as an introduction into radiation damage of human cells. The cell is composed of a nucleus surrounded by cytoplasm, which are both encased in an envelope of membranes. Although highly complex in structure, the nucleus cytoplasm is about 70% water. The most important part of the cell is the nucleus, usually an oval body near the center of the cell. Chemically the nucleus is very active. The normal growth of the cell is controlled by the nucleus, which initiates cell division and controls the repair of injured cells. Cytoplasm is more or less a colorless liquid that secretes enzymes and controls absorption and excretion of materials from the cell.

Many of the body cells have a very limited life span. Their functions require division of the cell at a certain stage in their life span. The parent cells pass their functions on to the daughter cells. As they divide, thread-like chromosomes begin to grow in the nucleus. The chromosome number is fixed for a given species. Genes arrange themselves in a line along the chromosomes, which determine hereditary characteristics. When the cell divides, the daughter cell receives a duplicate set of chromosomes from the parent, as well as identical genes. If the process is normal, no changes take place. However, when changes do take place they are designated as mutations. These mutations can then be passed on to the future daughter cells. Mutations are often the direct result of radiation.

The development of the whole body proceeds from a cellular process called mitosis, or cellular division. If the species is bisexual, as most are, the union of two cells (gametes)—the sperm from the male and egg, or ovum, from the female—produces an original cell from which the species will be produced.

The fertilized egg undergoes a number of divisions. Each newly divided cell all resemble each other in the embryonic stages. Changes in the structure of each cell, however, begin to take place with growth. The changes enable the cell to perform specialized functions. This process is known as differentiation. The result of this process is the development of two different cell types or lines. One of these lines is the germ line. The rest are called somatic. The germ line gives rise to either sperm in males or ova in females. The somatic lines develop into the tissues of the individual.

Only the genetic code in gametes (sperm and ova) can be transmitted to future generations of the species. Damage to somatic cells are confined to the individual. It should be noted that damage to a germ cell can be passed on to future generations. Thus, radiation damage can be passed on to future generations as well as an individual, which sheds some light on the long-term health effects of radiation.

Injuries to cells stem from a variety of agents and circumstances. The effects are the same regardless of the damaging agent. Ionizing radiation produces damage to cells in a very random manner, while other physical and chemical agents create a more orderly and predictable damage.

Radiation passing through living cells will ionize or excite at the molecular level within the cell structure. These changes affect the forces which bind the atoms together into molecules. When a molecule breaks apart, it triggers a chain of reactions within the cell. The broken molecular fragments are called radicals or ions and they are chemically unstable. Further effects are produced when the radicals and ions interact with other cell material. The role of each reaction plays an unsolved medical dilemma. The nucleus of the cell is the most affected portion and damage to cytoplasm may also cause severe adverse effects.

The total effect on the processes of cells is a function of the dose of radiation. The cell processes will be affected to varying degrees, up to and including death. Some damage to the cell may be repairable. This can be caused by action of the cell itself or by replacement of badly injured cells in a given tissue through mitosis of healthy cells. If the extent of damage to the organ is quite large, the organ may not be able to repair itself. Although many factors are important in assessing the total damage, it seems likely that most cell functions and associated structures are somewhat impaired by radiation.

A great deal of theoretical knowledge has been gained concerning the threshold dose below which no effects could be observed. At the present time, there is not enough empirical data to prove or disprove the existence of a threshold radiation dose number. For this reason, one should always assume that however small the amount of radiation, it will produce biological effects.

RADIO-SENSITIVITY

Since cells differ in both appearance and function, one might suspect that their individual response to radiation would also differ. This is true and that difference in response is known as radio-sensitivity of the cell.

Bergonie and Tibondeau were the first researchers to point out this difference in radio-sensitivity. They found that radio-sensitivity of a tissue is directly proportional to the reproductive capacity and varies inversely with the degree of cell differentiation. Since then, other factors have been found which affect the radio-sensitivity of a cell. Some of the other factors include the metabolic state of the cell, the state of cell division, and the state of nourishment. It has been found that to produce a given effect, the necessary radiation dose varies inversely as the relative radio-sensitivity of a given tissue.

Thus, cells are most active when reproducing at a rapid rate. Cells which have a high metabolic rate (rate of chemical changes in the cell) and those which are more nourished than others are more sensitive to radiation. There is some evidence that cells are more susceptible to radiation at certain stages of cell division than at other times. Cells that are more fully mature will be less damaged than cells which are still forming and growing. Bone marrow, lymphoid tissues, and the reproductive organs rank among the most radio-sensitive. Muscle and bone cells are the least radio-sensitive.

RADIATION DAMAGE

Damage to somatic cells is always limited to the individual, as previously discussed. Somatic effects include any and all types of damage to the individual. Damage to the germ cells (sperm and ova) is another matter. Germ cell damage can be passed on to the offspring of an individual. Thus, one may broadly classify damage as individual or hereditary. Heredity effects are those that can be transmitted to future generations. The term genetic damage, refers to effects caused by chromosome and/or gene mutations. The heredity effects are only applicable when damage effects the germ line, since only then can these effects be transmitted to a future generation.

SOMATIC EFFECTS

Effects on the somatic cells are usually expressed in terms of total body and partial body radiation impacts, with reference to individual organs. Certain organs of the body are more important than others and damage to these organs can lead to damage in less important organs. A number of physical factors are important in the determination of somatic effects. They include the following:

1. *Nature or type of radiation:* some types of radiation are more effective in producing damage.

2. *The absorbed dose:* this is a function of the energy absorbed per gram of tissue.

3. *Time distribution:* a potentially lethal dose given in a short time frame may not be lethal as it would be if protracted over a long period of time.

4. *Dose distribution:* the question must be addressed. Is the whole body involved or just a specific organ?

All of the factors stated above combine to create a matrix of effects that differ with different portions of the body. The age of an individual also becomes a factor, since children are more radio-sensitive than adults.

All of the effects can also be divided into early and latent effects. This is a very arbitrary classification, but it does serve to partition those effects noted in the first few weeks of exposure, compared to those that may occur over a period of years. The range of these effects, as well as the duration, depend upon the dose. The heavy doses create three forms of early and acute damage. They are listed below:

1. *Few thousand rad:* dose is fatal from a few minutes to hours. If the head receives the bulk of the severe dose, it breaks down the central nervous system and creates what is designated as central nervous system (CNS) death.

2. *500 - 2,000 rads:* symptoms may appear within hours. Death often occurs within a week or ten days. Damage to the lining of the intestinal tract is the most severe. This form of death is called gastronomical tract (GI) death. It is possible to survive this much radiation, only to succumb to lower doses at a later date.

3. *Less than 500 rad:* creates damage to the blood-forming organs. The mode of death is designated as bone marrow death. The first signs of sickness occur within the first few days of exposure. Severe damage occurs when the dose rate is greater than 200 rads. Any dose above 300 rad is usually fatal.

The above discussion is based upon the results of X ray and gamma ray radiation. As such, it is not applicable to convert to rems or other types of radiation.

The clinical effects which follow acute exposure to the total body are summarized in Table 7-4. Note the virtual absence of any symptoms in the range below a dose of 100 rems. However, some individuals would be expected to have mild symptoms in the range of 50-100 rems, because of the difference in tolerance or radiosensitivity. Below about 50 rems, no symptoms are expected to be noted in the first few weeks or months.

Death occurs in a larger fraction of the cases as the dose increases. If the dose becomes large enough, all cases of exposure result in death. In the range where survival is possible, the concept of median lethal dose (LD$_{50}$) is used. This expresses the dose at which 50% of those exposed will die. The best estimate places the LD$_{50}$ at 300-500 rad. This range would present severe symptoms. The reader should keep in mind that all of the above discussion is confined to short-term effects of severe doses of radiation. It fails to address the latent effects, which may range over a lifetime.

BLOOD AND BONE MARROW

The blood is composed of three cell types, which include the red cells (erythrocytes), the white cells (leukocytes), and platelets. All are suspended in a fluid called plasma. The red cells supply other body cells with food and oxygen and remove waste products. White cells aid in combatting infections, and the platelets aid in the clotting of blood at points of injury. Plasma is a viscous fluid which contains water, proteins, salts, and free ions. Blood contains about 45% red cells, ≈ 1% white cells and platelets, and 54% plasma.

The center of bones are the first to be affected by radiation. Even though there are subtypes of white cells, which differ in their radio-sensitivity, the net effect of radiation is to reduce the number of white cells. This lack of white cells is known as leukopenia. In cases of severe radiation, the platelets drop within a week. A few weeks later the red cells peak and anemia occurs. The loss of white cells affects the body's resistance to infection. The platelet number affects the clotting action, so that wounds will not heal. Anemia causes general weakness in individuals. Loss of red blood cells will not allow bone marrow to regenerate. Bone marrow damage is likely to be permanent.

Table 7-4. Summary of clinical effects of acute ionizing radiation.[3]

Range	Subclinical Range 0 - 100 rems	Treatable Range 100 - 1000 rems			Lethal Range > 1000 rems	
		100 - 200 rems	200 - 600 rems	600 - 1000 rems	1000- 5000 rems	> 5000 rems
		Clinical surveillance	Therapy effective	Therapy promising	Therapy palliative	
Incidence of vomiting	None	100 rems 5% 200 rems 50%	300 rems 100%	100%	100%	
Delay time		3 hours	2 hours	1 hour	30 minutes	
Leading Organ	None	Hematopoietic tissue			Gastro-intestinal tract	Central Nervous System
Characteristic signs	None	Moderate leukopenia	Severe leukopenia: hemorrhage, infection, purpura elation above 300 rems		Diarrhea; disturbance of electrolyte balance	Convulsions; tremor; alexia; lethargy
Critical Period of Exposure	None		4–6 weeks		5–14 days	1–48 hours
Therapy	Reassurance	Reassurance hematologic surveillance	Blood transfusion antibiotics	Consider bone marrow transplant	Maintenance of electrolyte balance	Sedatives
Prognosis	Excellent	Excellent	Good	Guarded	Hopeless	
Convalescent Period	None	Several weeks	1 - 12 months	Long	Death	
Incidence of Death	None	0 - 80%	80-100 %	90-100 %		
Death occurs within:		2 months		2 weeks	2 days	
Cause of Death		Hemorrhage; Infection		Circulatory collapse	Respiratory failure; brain edema	

LYMPHATIC SYSTEM

The lymphatic system is network of small tubes which permeate the body tissues. The fluid, lymph, has less protein than plasma and drains from tissues into the lymph system, sweeping up waste products. Along the course of the tubular system are oval-shaped glands (lymph nodes), which filter foreign substances from the lymph fluid. The purified lymph is then passed back into the blood stream. The spleen contains the largest mass of lymphatic tissue in the body. The spleen filters dead blood cells from the body and is the source of white cells. It also acts as a storage area for red blood cells. The lymph nodes show the first sign of hemorrhaging and infection after acute radiation. The spleen may exhibit weight loss and damage to lymphocytes (subtype of white blood cells).

DIGESTIVE TRACT

The digestive tract, or alimentary canal, consists of the mouth, pharynx, esophagus, stomach, and the small and large intestines. The canal may be as long as 30 feet in length in the adult male. The cells that line the walls of the intestine secrete a substance which acts on food to make absorption into the bloodstream possible. The stomach is the reservoir in which the chemical phases of digestion are initiated. The radio-sensitivities of many of the sections of the canal vary over a wide range. The small intestine is quite radio-sensitive, whereas the stomach and esophagus are much less sensitive to radiation.

The symptoms of damage to the canal are nausea and vomiting. The initial effects are impaired secretion and discontinued cell production. When cell breakdown occurs, the dead cells are released from the walls of the tract. This debris clutters up the intestine. The exposure of tissues under the surface may lead to ulcers. In fatal cases, infection, failure of food absorption, and dehydration from diarrhea are the causes of sickness and death.

REPRODUCTIVE ORGANS

Since the reproductive organs are the source of germ cells, damage to these cells can result in both somatic and hereditary effects. The response of germ cells to radiation differs slightly in the male and the female. If the dose is high enough, the effect will be sterility. Sterility requires a larger dose to the male compared to the female. In most cases, the dose required to produce permanent sterility is in the same region as fatal doses. Partial sterility can occur in man with doses around 150 rad. Germ cells that survive damage can transmit any genetic changes caused by radiation. For this reason, the total effect of radiation on the gonads may not be seen for several generations.

NERVOUS SYSTEM

The central nervous system, composed of the brain, spinal cord, and the peripheral nerves, acts to coordinate the body activity. The spinal cord and peripheral nerves are highly radio-resistant, but the brain is more sensitive than other portions of the central nervous system. The effect of radiation on the brain expresses itself as an alteration to functions, rather than structure for doses up to LD_{50}. At higher doses, brain damage may occur directly, or through a lack of blood supply, due to damage of the blood vessels.

THYROID GLAND

The thyroid gland is located at the base of the throat and secretes a hormone known as thyroxine, which aids the metabolism of the brain. The action of the thyroid seems to be closely connected with functions of the pituitary and adrenal glands. Thyroxine contains about 65% iodine, and is essential to growth and development. Damage to the thyroid or to the other two glands have marked effects on the rest of the body. The thyroid gland is radio-resistant from the standpoint of external radiation. It can be severely damaged if radio-iodine is inhaled, since iodine will concentrate in the thyroid. Damage causes a decrease in production of thyroxine, which produces a lower metabolic rate. Muscle tissue may fail to absorb enough oxygen and health can be badly impaired.

EYES

The lens of the eye is highly susceptible to irreversible radiation damage. The lens cells of the eye are not replaced by new growth. It is a latent effect and may not take place for years after the damage has occurred. Geologists who have worked in the uranium fields of the southwestern U.S. have been known to lose sight in the eye which they used to examine uranium specimens using a hand lens. Losing the lens can occur 30 years after exposure.

The retina is much less sensitive than the lens of the eye. When the cells of the lens become damaged, the cells lose their transparency. Acute effects in other portions of the eye occur only after high doses are received.

Neutrons are more effective than X rays in producing cataracts. Doses of X rays that exceed 500 rad will produce significant cataracts, but 200 rad of mixed gamma and neutrons have also created opacities of the lens. The susceptibility to cataract formation depends upon age. Radiation is more likely to cause cataracts in younger people.

LUNGS

The lungs are cone-shaped organs composed of small sacks called alveoli. As a person breathes, the air is directed down the trachea (wind pipe). Two large tubes branch into the lungs. Many small tubes branch from the bronchi to connect with the alveoli in the lungs. During breathing, each air sack is expanded and compressed by lung muscles, and is then filled and emptied of air. Air passes through the walls of the alveoli into tiny blood vessels called capillaries.

The damaging effects produced in the lung by radiation are the result of damage to the air sacks. The lungs are not normally effected by external radiation. As in the case of the thyroid, the greater hazard occurs from internal radiation from inhaled dust and vapors. Dr. T.D. Martonen, a biophysicist with the EPA's Health Effects Research Laboratory in North Carolina, recently reported that "... lung cancer can be initiated at a single-cell level and massive localized doses enhance the probability of lung cancer being induced."[6] This quote is taken out of context and refers to air pollution but is just as applicable to damage induced by alpha and beta particles.

LIVER AND GALL BLADDER

The liver is relatively radio-resistant compared to other organs. The liver, the largest gland in the body, secretes bile for digestion. The gall bladder stores and concentrates the bile secreted by the liver. When bile is needed, it passes into the intestine. External radiation is not too effective in causing damage to these organs. Most damage is caused by internal exposure from radioisotopes, which concentrate in the liver.

KIDNEYS

The kidneys help control the concentration and content of the blood by excreting water and waste products. The waste products pass from the kidneys through small tubes (ureters) into the bladder. Impairment of renal functions does not add to mortality in the case of total body radiation. Damage to the kidney is indicated by an increase in amino acids in the urine. The appearance of blood in the urine is an indication of severe renal damage.

CIRCULATORY SYSTEM

The heart and blood vessel system are damaged seriously only for very high doses of radiation.

SKIN

The degree of skin damage varies with the dose and the species of animal. Skin is easily damaged, but has a tremendous capacity for repair. Various structures of the skin show quite differing sensitivities. The damage seems to be greater for less penetrating radiation. Slight damage to the skin may result in a reddening (erythema). Skin cancer is a latent effect of chronic radiation in relatively high dose rates.

HAIR

Radiation can lead to temporary baldness, which may last for a few weeks. The hair begins to return, often in a new color.

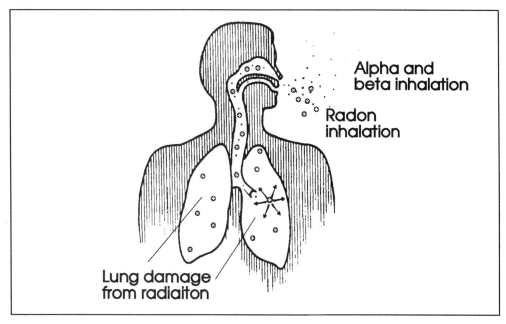

Figure 7-3. Lungs of an individual contaminated by NORM.[4]

BONES

Bone is composed of living cells, which are distributed in a matrix of fibers and bone salts. Although the marrow of the bone is radio-sensitive, the bone cells, fibers, and salts are relatively radio-resistant. When radioisotopes such as strontium or plutonium are internally deposited in the bone marrow or bone tissue, then great damage may occur. These are also latent effects, which may take years to surface.

LIFE SPAN

Information on life-shortening effects in man is still inadequately known. The effects of long-term, low-level radiation on longevity cannot be predicted. Much more data must be obtained before any conclusions can be drawn.

SUMMARY

The sum total of our knowledge on low-level radiation that can create ill health effects can be outlined in a few sentences.

1. If a large group of people are exposed to significant levels of radiation, more cancer will develop among that group as compared to a similar group that was not exposed to radiation.

2. The cancers developed will not be unique in types or characteristics as opposed to nonexposed cancerous cells.

3. It is not known what dose is required to induce cancer.

4. It is not known why two persons receiving the same dose may not have similar future health effects.

5. If 100,000 people are exposed to 100 millirems each year of their life, about 520 will die of cancer.

6. Based on current statistics, about 16,800 people out of 100,000 will develop and die of naturally induced cancer.

7. Statistical information often has numerous variables which may not apply to any particular inquiry.

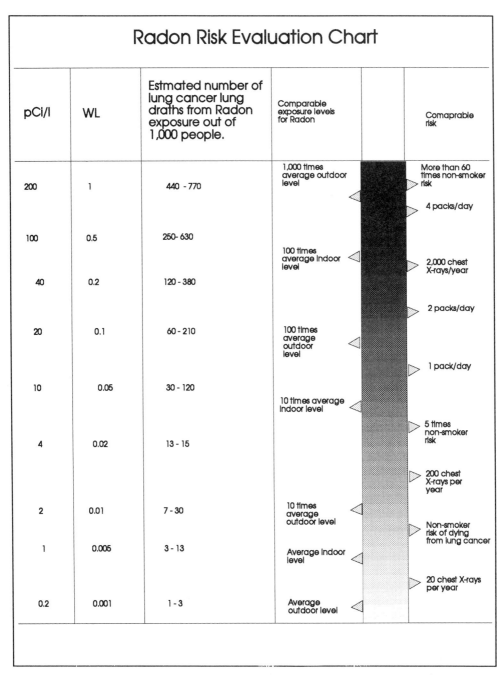

Radon Risk Evaluation Chart

pCi/l	WL	Estmated number of lung cancer lung draths from Radon exposure out of 1,000 people.	Comparable exposure levels for Radon		Comaprable risk
200	1	440 - 770	1,000 times average outdoor level		More than 60 times non-smoker risk
					4 packs/day
100	0.5	250- 630			
			100 times average indoor level		2,000 chest X-rays/year
40	0.2	120 - 380			
					2 packs/day
20	0.1	60 - 210	100 times average outdoor level		
					1 pack/day
10	0.05	30 - 120			
			10 times average indoor level		
4	0.02	13 - 15			5 times non-smoker risk
					200 chest X-rays per year
2	0.01	7 - 30	10 times average outdoor level		
					Non-smoker risk of dying from lung cancer
1	0.005	3 - 13	Average indoor level		
					20 chest X-rays per year
0.2	0.001	1 - 3	Average outdoor level		

Figure 7-4. Radon risk evaluation chart comparing various risks to annual radon exposures.[4]

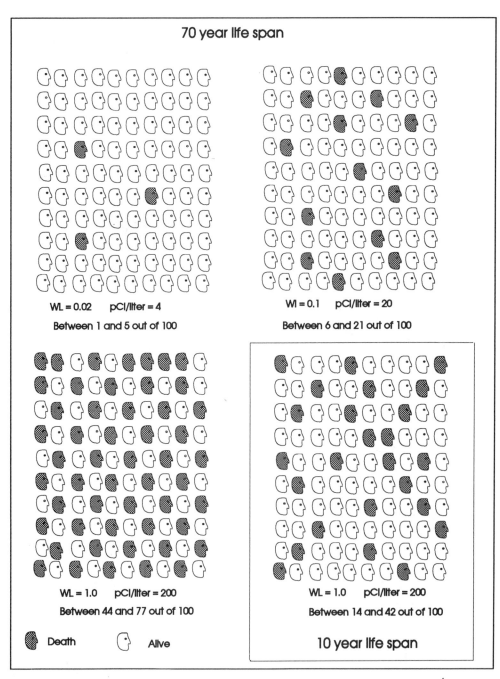

Figure 7-5. Pictorial chart displaying radon risk for various Working Levels and exposure rates.[4]

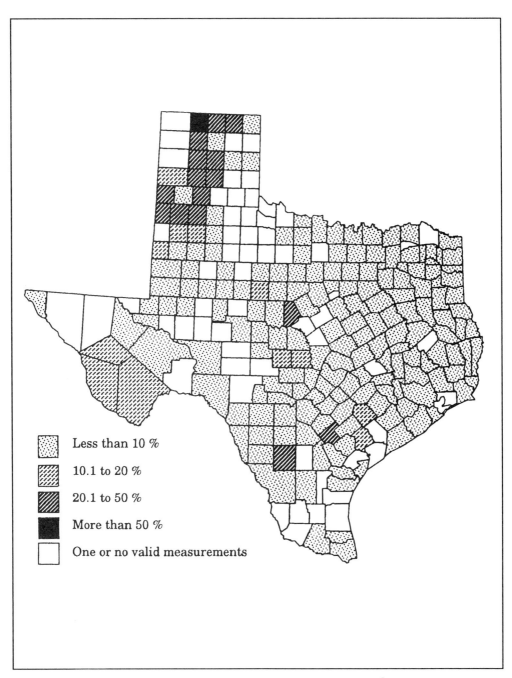

Figure 7-6. Average radon levels of exposure in Texas to greater than 4 pCi/liter.[5]

Less than 10 %

10.1 to 20 %

20.1 to 50 %

More than 50 %

One or no valid measurements

CHAPTER 7
REFERENCES

1. Wilson, R., and Hendricks, W. G. "NORM: Occurrence, Handling and Regulations," Hazardous Materials Control Resource Institute Conference, New Orleans, 1992.

2. Suntrac Services, Inc., Naturally Occurring Radioactive Materials Seminar, University of Houston Institute for Environmental Management, 1992.

3. *The Effects of Nuclear Weapons*, U.S. Government Printing Office, Washington 25 D.C.

4. *A Citizens Guide to Radon*, U.S.E.P.A., OPA-86-004, Office of Air and Radiation, 1986.

5. Texas Department of Health, *The Texas Radon Indoor Radon Survey*, 1992.

6. "News Update," magazine, Environmental Protection, p. 8, April, 1993.

7. Moghissi, A.A., and Carter, M.W., *Evaluation of Public Health Implications of Radioluminous Materials, Radioactivity in Consumer Products*, NUREG, CP-0001, USNRC, August, 1978.

8. Hallenbeck, W.H. and Cunnigham, K.M., *Quantitative Risk Assessment for Environmental and Occupational Health*, Lewis Publishers, Inc., 1988.

CHAPTER 8

OVERVIEW OF NORM REGULATIONS

I t was stated in 54 A.D. by that great Roman philosopher, Gluteus Maximus, that "keeping up with regulations was akin to nailing Jello to a tree." Gluteus went on to state that regulations inspired an inverse relationship with respect to comprehension.

Regulations evolve from the regulatory agencies through laws set up by federal and state legislative bodies. The regulatory agencies are given the mandate to promulgate regulations through enforcement. Sometimes, the regulations created by agencies are then run back through the legislative bodies to be "codified" as law. Sometimes, they are modified or new regulations are added into the code at this juncture. A regulation to the public and business is a "de facto" law. It is often difficult to keep track of all the new proposed regulations and policies. Regulations are usually published in the Federal Register or a state register for a period of time to receive comments. After about 60-90 days or some indefinite period, it is once again published in a register, and becomes effective and enforceable as of a certain date. Policy and guideline statements by regulatory agencies are often a precursor step to new regulations. The policy statement is another vehicle to receive comments prior to formal promulgation of a regulation.

NORM REGULATIONS

The states have taken the lead in creating NORM regulations and laws, due to the absence of any federal guidelines for low-level radioactive waste. The *Nuclear Regulatory Commission* was set up to implement and enforce the *Atomic Energy Act*, which primarily addressed the atomic energy industry. NORM was not an issue at the time, and it simply was ignored. The states began to fill this vacuum. It is predicted that the federal regulatory agencies will not continue to ignore the NORM problem in the near future.

A person responsible for regulatory compliance within a company must first identify the waste streams and production sources that may create a NORM waste. The question then becomes, "Is the handling, treatment, and disposal of NORM material regulated?" There are four possible answers:[1]

   1. The material is regulated as a radioactive waste.

2. The material is regulated as a waste substance, with or without regard to the fact that it contains radioactivity.

3. There are limits imposed upon handling the materials by virtue of the common law.

4. The material is not regulated.

NORM will be regulated in all of the states that have the material associated with any industrial process. It will be regulated to some degree under the *Safe Drinking Water Act*. It is presently regulated in the states of Arkansas, Louisiana, and Mississippi. It will soon be regulated in the following states:[2]

Alabama	Kentucky
Michigan	Alaska
Illinois	North Dakota
New Mexico	Kansas
Oklahoma	Ohio
Texas	

Many other state and federal regulatory agencies will undoubtedly be added to this list in the near future. There will be confusion and variations, but there will be regulation. The states want to maintain *primacy*, which means they must equal or exceed any federal laws or regulations. The state agencies do not generally want the federal regulatory agencies governing in their domain, thus the states are moving ahead of the federal agencies concerning some environmental issues, and NORM is a prime example.

One of the most contentious issues in the evolving NORM state regulations is over the levels of exposure or risk, and what threshold level should be regulated. Another group of issues that presents more confusion are the units of measurement that apply to these threshold levels. A third common issue is disposal. At present it is very difficult to dispose of NORM waste, if not economically impossible in some instances. The states and the federal regulatory agencies are wrestling with this issue and they are making some progress. For example, the *Minerals Management Service* (MMS) will now allow *Class II injection wells* to receive NORM waste with a permit.[3] The *Louisiana Department of Environmental Quality (LDEQ)* is looking for a suitable landfill site in Louisiana. The Texas Department of Public Health has handed the disposal problem to another agency, the *Texas Water Commission (TWC)*, which will soon become part of the Texas Natural Resource Conservation Commission. The *American Petroleum Institute (API)* has made recommendations for the disposal of NORM.[4] Some of these issues will continue to boil over the next few years, until they are resolved. It is a rather normal and expected process, as more people increase their awareness. Disposal will probably be the most controversial issue. Table 8-1 depicts some of the threshold levels now being proposed and/or enacted. The table is an example of the confusion over what units to use and what to measure.

Louisiana changed its field measurement surveys due to the original confusion and made all pieces of equipment or NORM-radioactive materials subject to license, if they emitted more than 25 µR/hour above background. This would equate to about 33 µR/hour for onshore Louisiana. Texas has proposed a licensing limit of 50 µR/hour including background. The reader should note that these references are to gamma, alpha, and beta radiation emissions. It is expected that most states will adopt regulations that fall within the 25-50 µR/hour for total radiation count to trigger permitting and licensing with annual fees for NORM sites. Some of the proposed regulations express "50 µR/hour" in terms of "50 microroentgens per hour."

Table 8-1. Acceptable surface contamination levels for NORM.

Radionuclide	Average contamination level per 100 cm^2.	Maximum contamination level per 100 cm^2.	Removable per 100 cm^2 of radioactive material via a wipe test.
U^{natural}, U^{235}, U^{238} and immediate decay products (Th^{234}, U^{234})	5,000 dpm of alpha particles per 100 cm^2	15,000 dpm of alpha particles per 100 cm^2	1,000 dpm per 100 cm^2
Ra^{226}, Ra^{228}, Th^{natural}, Th^{228}, Th^{230}, Th^{232}, Pa^{231}, Ac^{227}	1,000 dpm per 100 cm^2	3,000 dpm per 100 cm^2	200 dpm per 100 cm^2
Beta-gamma emitters including Pb^{210}, (nuclides with decay modes other than alpha emission or spontaneous fission) except others noted above.	5,000 dpm beta-gamma per 100 cm^2	15,000 dpm beta-gamma per 100 cm^2	1,000 dpm beta-gamma per 100 cm^2

HISTORY OF NORM REGULATIONS IN LOUISIANA

The state of Louisiana has been a leader in the field of NORM regulation. It developed early in Louisiana due to the concentration of oil and gas production with associated NORM problems. The Louisiana Department of Environmental Quality (LDEQ) on November 8, 1988, issued a memorandum of "Guidance for Dealing

Radium Scale and Associated Problems."[7] The memorandum listed concerns and made recommendations for an action plan for dealing with the NORM problem found in the oil and gas fields. It also contained an interim policy, which established radiation procedures with regard to handling, storing, and disposing of NORM. The Radiation Protection Division of the LDEQ amended the Louisiana Radiation Regulations on September 20, 1989, and added Chapter 14, entitled "Regulation and Licensing of Naturally Occurring Radioactive Material (NORM)" to LAC 33:XV. This new law established "Radiation Safety Requirements" for possession, use, transfer, and disposal of NORM. It also established a fee category of $100.00 per annum for licensing of NORM sites. Section 1410 provided that "Each General License shall establish Written Procedures to ensure worker protection and for the survey of equipment and components to ensure that the levels in Section 1410 are not exceeded." Under this amendment to the Louisiana Radiation Regulations, "any person who engages in the extraction, mining, beneficiating, processing, use, transfer, or disposal of NORM, including scale deposits in tubulars and equipment and to soil contamination by the cleaning of scale deposits" is subject to the regulation. It established a ground contamination limit for Ra^{226} of 5 pCi/gram above background, averaged over the first 15 cm of soil, and 15 pCi/gram above background, averaged over a 100 m^2 area. It should be noted that "pCi/gram" can only be measured in a laboratory. It also changed the contamination limits for Ra^{226} for exemptions of tubulars and equipment from 30 pCi/gram to an exposure level of 50 μR/hour, including background. The LDEQ felt this change was necessary in order to facilitate survey work in the field, allowing companies a radiation exposure level that could be measured with field instruments.

The LDEQ issued a regulatory guideline on March 12, 1990, for performing surveys of land and equipment. This guideline set a deadline of July 1, 1990, for submitting confirmatory surveys of all potential NORM sites. It provided procedures for sampling in the field and set protocols for lab analysis. This was followed on October 25, 1990 by a memorandum stating that companies which had not complied were subject to civil penalties. A limit was set for reporting purposes at 50 µR/hour. This limit has now been lowered to 25 µR/hour. The LDEQ estimates that as many as 20,000 sites may be contaminated with NORM in Louisiana.

Table 8-2. Practical Quantitation Levels (PQL) for radionuclide contaminants.[5]

Radionuclide Contaminant	PQL (pCi/liter)
Radium ^{226}	5
Radium ^{228}	5
Uranium (natural)	5
Radon ^{222}	300
Gross alpha emitters	15
Gross beta emitters	30
Radioactive cesium ^{134}	10
Radioactive cesium ^{137}	10
Radioactive iodine	20
Radioactive strontium ^{89}	5
Radioactive strontium ^{90}	5
Tritium	1,200

ENVIRONMENTAL PROTECTION AGENCY AND NORM

The Environmental Protection Agency proposed NORM regulations concerning National Primary Drinking Water Regulations and Radionuclides in Drinking Water in the Federal Register on July 18, 1991.[6] This proposal allows water supply *Practical Quantitation Levels (PQLs)* for radionuclides, as shown in Table 8-2.

Some workers have expressed the opinion that the above referenced standards may become incorporated into state regulations on NORM in ground water.

The *Federal Code of Regulations (CFR)* has fifty titles for various federal

Table 8-3. Some CFR references for the regulation of radiation. Title 10=Energy; Title 29=Labor; Title 40=Protection of Environment.

CFR Title and Part	Subjects Concerned with Radiation
10 CFR Part 20	Radiation protection standards
10 CFR Parts 30-35	Byproduct material
10 CFR Part 39	Well logging licenses and radiation safety requirements
10 CFR Part 50	Nuclear Regulatory Commission domestic licensing of production and utilization facilities
10 CFR Part 70	Special nuclear material, domestic licensing
10 CFR Part 960	Nuclear waste repositories, general guidelines for recommendations of sites
29 CFR Part 1910	Occupational safety and health standards
40 CFR Parts 141-142	National primary drinking water standards
40 CFR Part 190	Environmental radiation protection standards for nuclear power operations
40 CFR Part 191	Environmental radiation protection standards for management and disposal of spent nuclear fuel
40 CFR Part 192	Health and environmental protection for uranium and thorium mill tailings

agencies, which often overlap. The following table lists a few salient references to radiation regulations. Some of these referenced sections may become resting places for regulations set in stone.

REGULATORY OVERVIEW

Massive public and learned ignorance exists concerning the effects of NORM. This ignorance is driven by fear and confusion over how much radiation is too much. There is scientific ignorance over the acceptable levels of NORM, and there are very few studies relating to the dangers of NORM. It is an emerging area of knowledge, and only three states have taken concrete action to deal with NORM on a regulatory basis at the present time. The federal statutes and regulations often overlap and do not specifically address the issue of NORM. Thus, when confronted by the fact that managers and owners must deal with NORM, the following guidelines are suggested.[1]

1. *Do not ignore the situation*: The public will expect you to be knowledgeable on how to handle the material in a competent and prudent manner. Despite the lack of quantitative data, existing and pending regulations will require industry to comply with stringent NORM- contaminated waste disposal procedures. Compliance will also require increasing employee personal protective equipment and training. Industry will bear the compliance cost, whether or not the procedures are absolutely necessary in all instances.

2. *Do not focus on the radioactive material aspects of the substance by considering only that one characteristic*: You must also consider whether or not it is toxic or dangerous to handle in other contexts, and indeed, the "other contexts" may drive your decisions on how to handle the material responsibly.

3. *Be prepared to communicate*: Communicate on a reasonable basis. Do not understate or overstate the risks and problems associated with the material.

4. *Keep up with developing regulations*: This must be done for two major reasons. First, regulators need your input from the initial moment of the idea to regulate. Second, if you are dealing with NORM, you should consider all of the ramifications that relate to your business, producing property, or services to maintain their value. A baseline proactive stance will save a lot of time, money, and agony over the long term.

5. Engage the services of an environmental consultant: A baseline environmental audit of your facilities will go a long way toward the goal of legal compliance. This will serve the purpose of protecting your future by virtue of the common law.

EXISTING FEDERAL LAWS AND REGULATIONS

The history and development of Louisiana NORM regulations have already been discussed. They are important because they will be used as a model for other states to modify and follow. It must be understood that other federal and existing state laws and regulations do overlap into the NORM area, in one way or another. An attempt will be made to present a broad overview of these regulations and how they might or might not fit into the NORM scheme.

The broadest law, encompassing the governing of radioactive material, is the Atomic Energy Act. This act has been amended from time to time and grants authority to the Nuclear Regulatory Commission to regulate certain radioactive substances. The focus of attention has been on the byproduct material, source material, and special nuclear material. The act has recently been amended to include high-level radioactive waste, spent nuclear fuel, and transuranic wastes. NORM does not fit into any of these categories, as long as the concentration of the radioactive material is below a certain regulated threshold. For example, uranium and thorium in concentrations greater than 1/20% by weight are defined as source materials under the Atomic Energy Act. The act authorizes the NRC and state governors to enter into agreements or contracts. These contracts or agreements allow the individual state to oversee a program regulating certain types of radioactive material that would otherwise be completely administered by the NRC. Some of the materials, such as high-level wastes, are incapable of being regulated by statute as radioactive substances.

A state desiring to regulate the radioactive substances must devise a state program equivalent to the federal program regulating the same substances.

There are other Federal Acts which also may apply to NORM handling, as the result of NORM's radioactivity. Two of these include CERCLA (Superfund Act), and the Clean Air Act.

States which choose to regulate radioactive substances may have laws containing broader definitions than are present in the Atomic Energy Act. For example, in Texas a "radioactive material" is any substance, whether naturally occurring or artificially produced, that emits radiation spontaneously. It is broad enough to include the K^{40} in the bones of human citizens of Texas. It certainly includes NORM and X ray machines, neither of which is regulated by the NRC.

If definitions are that broad, and if the substance is, in fact, not exclusively regulated by the NRC or other federal agencies as a radioactive substance, then the opportunity exists for independent state regulation. Concentrations of NORM below the threshold limit of 2,000 pCi/gram are not regulated by the NRC.

The EPA has also been considering the regulation of NORM with a dividing line at the 2,000 pCi/gram for proposed regulatory purposes. The EPA has publicly stated that NORM below 2,000 pCi/gram can be hazardous to health depending upon the circumstances. The concern was expressed in a policy statement quoted below.

"In view of the uncertainties involved in risk assessment at low doses...the Commission finds that the average dose to individuals in the critical group, should be less than 10 mrem/year...for each exempted practice. In addition, an interim dose criterion of 1 mrem/year...average dose for individuals in the critical group will be applied to those practices involving widespread distribution of radioactive material...until the commission gains more experience with the potential for individual exposures from multiple licensed and exempted practices."[8]

It can readily be ascertained that a great deal of uncertainty exists around the threshold limit of safety concerning radioactive materials. In an effort to draw lines around the regulations, there are tables that may be or have been adopted by rule, which describe isotopes and their emissions, as shown previously. Some are confusing and only partially cover the broad spectrum of radioactivity. Some are very specific. Since the federal government has not specifically regulated NORM at low levels of radioactivity, the states have been free to promulgate regulations. This may not last, and the federal agencies may soon step up their efforts to regulate NORM as they watch developments within the states. This may lead to modifications within existing regulations, as well as prompting new regulations.

The group that has been at the forefront of regulation is the *Conference of Radiation Program Control Directors (CRPCD)*, made up of personnel from state radiation control programs in states, that exercise authority from their agreements with the NRC. This group has had an effort underway for several years to regulate *Naturally Occurring and Accelerator Produced Radioactive Materials (NARM)*. Most of NARM is now in the arena of NORM. The progeny of the committee has been *Part N*. This name is nothing more than an identifier for a proposal which would be considered by the state program directors and offered to the NRC as a proposal for adoption to be imposed upon the agreement states, so that nationwide uniform regu-

Table 8-4. Field survey threshold licensing trigger points for regulated states.

State	Field Survey Parameters
Louisiana	25 µR/hour above background
Mississippi	25 µR/hour above background
Arkansas	25 µR/hour above background expected, however at present it is set at 5 pCi/gram 15 cm or above and 50 pCi/gram below 15 cm of soil

lation can be promulgated.

Recently, efforts on a national scale have been drifting. The field survey parameters for the three regulated states are listed in Table 8-4.

Texas, as of April, 1993, has proposed a threshold limit that states "equipment containing NORM shall not exceed a maximum radiation exposure level of 50 µR/hour including background."[5] Texas regulations became effective on July 1, 1993, administered by the Texas Department of Health. Disposal is now administered by the Texas Railroad Commission and the Texas Natural Resource Conservation Commission (the former Texas Water Commission).

CHAPTER 8
REFERENCES

1. Wilson, R., Hendrick, W.G., "NORM: Occurrences, Handling and Regulation," Hazardous Materials Control Resources Institute Conference, 1992.

2. Gray, P.R., "Regulations for the Control of NORM," Transactions, SPE/EPA Environmental Exploration and Production Conference, 1993.

3. Rutherford, G.J., Richardson, G.E., "Disposal of Naturally Occurring Radioactive Material from Operations on Federal Leases in the Gulf of Mexico," SPE 25940, SPE/EPA Exploration and Production Conference, p.101-110, 1993.

4. American Petroleum Institute, Bul. E 2: *Management of Naturally Occurring Radioactive Materials (NORM) in Oil and Gas Production*, 1992.

5. Texas Department of Health, *Texas Regulations for Control of Radiation*, TRCR Part 46, Texas Register, March 23, 1992.

6. EPA, 40 CFR Parts 141, 142, RIN 2040-AA 94, "National Primary Drinking Water Regulations; Radionuclides," *Federal Register*, V 56, No. 138, p.33050-33127, July 18, 1991.

7. LDEQ, *Environmental Regulatory Guide Part XV*, "Radiation Protection," June, 1992.

8. May, H.D., Statement made before the Committee of Natural Resources, Texas State Senate, EPA, October 5, 1989.

RADIATION PROTECTION

INTRODUCTION

The first step to protection is awareness. This can only be achieved through training and the use of radiation detection equipment. Remember that radiation cannot be detected through smell, hearing, or visual observation. It is beyond the human senses, therefore the only way to make certain that workers are protected from radiation is to know where it might occur and understand the methods of detection.

DETECTING GAMMA RADIATION

Gamma radiation is an energy that requires the shielding of lead in order to be attenuated. The health hazards of gamma radiation occur when it travels through the body. Field survey meters measure gamma radiation in μrem/hour (roentgen equivalent-man) to calculate dose rates.

Gamma contamination in a piece of equipment or tubular could be detected with a survey meter if the following factors are present:

1. The survey meter is used with a gamma scintillation probe.

2. The contamination has a strong gamma-emitting isotope.

3. The amount of gamma activity is high enough to penetrate the wall of the equipment, tubular, soil or water, but all of the radiation is not attenuated.

4. The distance between the contamination and the survey meter is a very important factor. It should be noted that gamma radiation sensitivity is inversely proportional to distance.

DETECTING BETA RADIATION

Beta radiation is an energized electron particulate that requires shielding such as aluminum in order to be attenuated. The health hazards of beta radiation occur when it enters the body by ingestion, absorption, or inhalation. Field survey meters measure beta radiation in *counts per minute* (CPM) to calculate the radioactivity of the contamination.

Beta contamination in a piece of equipment, tubular, soil or some other material, can be detected if the following factors are present:

1. The survey meter is used with an end window or "pancake" probe with a

known efficiency for beta radiation.

2. Obstacles between the detector and the contamination create attenuation equivalent to a piece of aluminum or more dense material.

3. The distance between the contamination and the survey meter is an important factor. Note, that all readings should be taken with survey meters as close as possible to the contaminated piece of equipment. Detection is rare when detection equipment is placed more than 1 ft away from the contaminated material.

DETECTING ALPHA RADIATION

Alpha radiation is a high energy particulate that is generally a contaminant when it becomes airborne. The health hazards of alpha radiation occur when an alpha particle enters the body through ingestion, inhalation, or absorption. Field survey meters measure alpha radiation in *counts per minute* (*CPM*) to calculate the activity of the contamination.

Alpha contamination in a piece of equipment or other material can be detected with a field survey meter if the following conditions are met:

1. The survey meter is used with an end window or "pancake" probe with a known efficiency for alpha radiation.

2. Obstacles between the detector and the contamination create attenuation beyond the thickness of a piece of paper.

3. The probe is placed as closely as possible to the source of contamination. Alpha particles only emit energy for a few inches and are rarely detected beyond a distance of more than 2 in. It is obvious that a few inches of soil will attenuate alpha radiation.

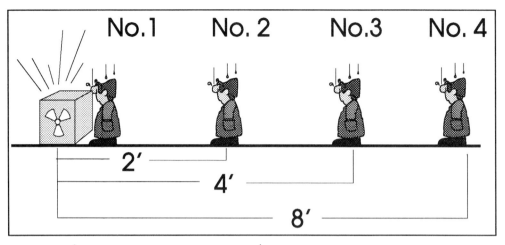

Figure 9-1. Inverse Square Law of Radiation Detection.[1]

BASIC CALCULATIONS FOR GAMMA DOSE RATES

The inverse square law is used to determine dose rates for known distances. The example problem and formula given below apply to point sources of radiation. Dose rates cannot be calculated accurately at a distance greater than about ten times the size of the radiation source. For example, if a piece of equipment were 10 ft in diameter and completely contaminated, dose rates could be calculated out to approximately 100 ft.

The cartoon drawing depicted in Figure 9-1, demonstrates the *Inverse Square Law of Radiation Detection*. The figure is to be used with an example calculation.

EXAMPLE: An external gamma survey of a piece of equipment is being conducted by a person as shown in Figure 9-1. The reading on the dial, taken 1 ft away from the piece of equipment, is 1,000 µR/hour for Person No. 1, standing 1 ft away. No. 2 is standing 2 ft away from the piece of equipment. Person No. 3 is standing 4 ft away. Person No. 4 is standing 8 ft away from the equipment. Assume that all

Table 9-1. Solutions to inverse square law.

Person	Distance	Dose Rate	Dose per 6 minutes
No. 1	1 ft	1,000 μR/hr	100 μR
No. 2	2 ft	250 μR/hr	25 μR
No. 3	4 ft	62.5 μR/hr	6.25 μR
No. 4	8 ft	15.5 μR/hr	1.5 μR

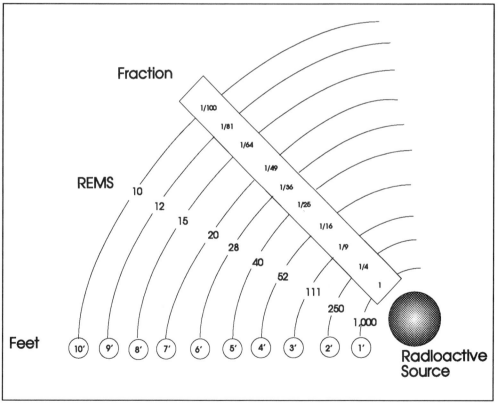

Figure 9-2. The relationship of distance to a radioactive source.[5]

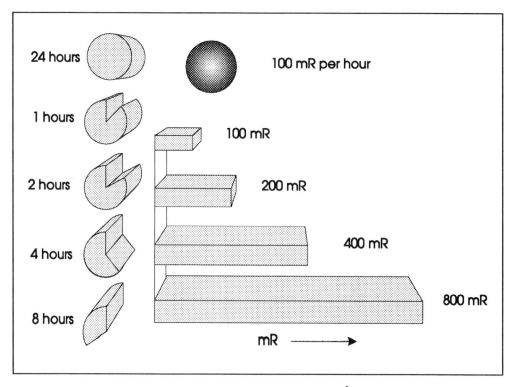

Figure 9-3. The relationship of time of exposure to cumulative exposure.[5]

survey readings were above background. Also, assume the persons stood at their stations for 6 min. Calculate the dose rate for each person.

Solution: 6 minutes / 60 minutes/hour = 0.1 hour

Inverse Square Law : $D_x = D_o (r_o/r)^2$

Where: D_x = Dose rate at X feet

D_o = Known dose rate

r_o = distance of known dose rate

r = distance of Y person.

Person No. 1: $D_1 = 1,000 \ \mu R/hour \times (1/1)2 = 1,000 \ \mu R/hour$

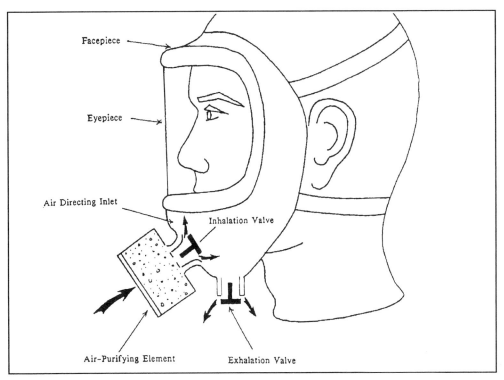

Figure 9-4. Full-face mask high efficiency respirator.

Face mask respirators are necessary when cleaning tubing and equipment with or without water. Tyvex suits, with long sleeves and boots and neoprene gloves, are also recommended. Alpha and beta particles, which can be airborne particulates, can be absorbed through the skin as well as inhaled or ingested. If equipment is to be broken down in the field and replaced or cleaned, it is recommended that workers take the necessary precautions of wearing personal protective equipment. Survey workers need not take all of the above precautions, unless they perceive a danger of exposure to high levels of alpha and beta dust and/or gamma radiation.

Table 9-1, shows the results of all four person's dose rate and dose for 6 minutes.

TIME, DISTANCE, AND SHIELDING

The methods to reduce radiation exposure to levels that are reasonably achievable are time, distance, and shielding. Time of exposure is critical to workers. The less time, the less chance of damage. This is just a statistical number and should not be construed to fit all cases and all individuals. The inverse square law demonstrates that distance is a prime factor in worker safety, as demonstrated in the example. Shielding is always a good precaution to take, whenever feasible.

Genetics of the individual can also be a large factor in doses that might be considered low by existing standards or by standards that may be promulgated in the future. This is an unknown area of concern and much depends upon the radio-sensitivity of the individual. Precaution and reasonable safety measures should always be practiced when working with radiation. Figure 9-2 depicts the concepts of the distance factor as it relates to a radioactive source. Note the lower fractional dosage rates with increasing distance and rem reduction.

Figure 9-3 diagrammatically depicts the relationship of exposure time to the cumulative effects of radiation over an eight-hour period. The source is emitting at a rate of 100 mrem/hour.

Personal Protective Equipment (PPE)

Personal Protective Equipment (PPE) should be worn when working with radioactive material, particularly if it is dry material that could become airborne. NIOSH (National Institute for Occupational Safety and Health) recommends full-face mask respirators, with high efficiency particulate filters capable of filtering dust down to 0.3 microns dioctyl phthalate particle with 99.97% efficiency. Figure 9-4 illustrates a full-face mask filter respirator.

Table 9-2. Representative radioactive isotopes at four levels of toxicity.[3]

Toxicity Class	Representative Radioactive Isotopes
Class I (Very highly toxic)	Sr^{90}, Y^{90}, Pb^{210}, Bi^{210}, Po^{210}, At^{211}, Ra^{226}, Ac^{227}, U^{233}, Pu^{239}, Am^{241}, Cm^{242}
Class II (Highly toxic)	Ca^{45}, Fe^{59}, Y^{91}, Ru^{106}, Rh^{106}, Ba^{140}, La^{140}, Ce^{144}, Pr^{144}, Sm^{151}, Eu^{154}, Tm^{170}, Th^{234}, Pa^{234}, natural uranium
Class III (Moderately toxic)	Na^{22}, Na^{24}, P^{32}, S^{35}, Cl^{36}, K^{42}, Sc^{46}, Sc^{48}, V^{48}, Mn^{52}, Mn^{54}, Mn^{56}, Fe^{55}, Co^{58}, Co^{60}, Ni^{59}, Cu^{64}, Zn^{65}, Ga^{72}, As^{74}, As^{76}, Br^{82}, Rb^{86}, Zr^{95}, Nb^{95}, Mo^{99}, Tc^{98}, Rh^{105}, Pd^{103}, Rh^{103}, Ag^{105}, Ag^{111}, Cd^{109}, Ag^{109}, Sn^{113}, Te^{127}, Te^{129}, I^{132}, Cs^{137}, Ba^{137}, La^{140}, Pr^{143}, Pm^{147}, Ho^{166}, Lu^{177}, Ta^{182}, W^{181}, Re^{183}, Ir^{190}, Ir^{192}, Pt^{191}, Pt^{193}, Au^{198}, Au^{199}, Ti^{200}, Ti^{202}, Ti^{204}, Pb^{203}
Class IV (Slightly toxic)	H^{3}, Be^{7}, C^{14}, F^{18}, Cr^{51}, Ge^{71}, Ti^{210}

CLASSES OF RELATIVE RADIOACTIVE TOXICITY

There are four classes of radioactive toxicity levels as designated by the Atomic Energy Agency. These are usually labeled and often; it is good safety practice to train workers and label equipment with proper classifications. The following classes are noted:

Class I   Very highly toxic

Class II   Highly toxic

Class III Moderately toxic

Class IV   Slightly toxic

Table 9-3. Shipping label limits as provided by the DOT, 4a/CFR 172.403.

Warning Labels	Maximum dose rate at the surface of the package	Maximum dose rate at 3 ft from the package
Radioactive White I	0.5 mR/hr	Not specified
Radioactive Yellow II	50 mR/hr	1.0 mR/hr
Radioactive Yellow III	No maximum limit above 50 mR/hr	No maximum limit

NORM might fall into all four classes of toxicity, but generally will be concentrated in the Class III and Class IV categories. Toxicity classes and representative radioactive isotopes for each class are represented in table 9–2.

WARNING LABELS

The warning labels that must be provided for shipping purposes by Department of Transportation (DOT) are given in table 9–3.

GENERAL PROTECTIVE GUIDELINES

The following guidelines are recommended when dealing with NORM. These are standards promulgated by safe practices in industrial hygiene.

1. Employees and/or contractors should be advised when the presence of NORM is detected, and specified written procedures should then be followed.

2. Gloves and other personal protective clothing should be worn when dismantling contaminate equipment in order to prevent skin, eye, ingestion, and inhalation contact.

3. Prior to eating, drinking, or smoking, hands should be washed thoroughly, as well as at the end of each working day.

4. A high-efficiency, full-face mask respirator should be worn whenever grinding, drilling, polishing, welding, brazing, metalizing, or blasting on or near contaminated equipment for protection against airborne radiative matter. Tyvex suits and boots should also be worn to avoid absorption into the skin surfaces.

5. The following precautions should apply when scale or sludge contamination removal activities are conducted:

    a. Removal activities should be conducted in well-ventilated areas where access has been restricted.

    b. To minimize the generation of airborne particulates, scale and other types of NORM should be handled in a wet state.

    c. Floor covers, such as plastic sheeting, should be utilized to minimize the spread of contamination.

    d. High-efficiency, air-purifying full-face masks, Tyvex coveralls, and neoprene gloves should be worn with disposable boots as personal protective equipment.

    e. Proper training should be provided to the workers in the use and testing of the personal protective equipment, following NIOSH guidelines.

The following table shows the maximum permissible dose limit equivalent for occupational exposure as referenced to NRC.

Table 9-4. Maximum permissible dose equivalent for occupational exposures.[5]

Combined Occupational Exposure	Roentgens Equivalent-Man (REMs)
Prospective annual limit	5 rems in any 1 year
Retrospective annual limit	10-15 rems in any 1 year
Long-term accumulation	(Nx18) x 15 rems, where N = age in years
Skin	15 rems in any 1 year
Hands	75 rems in any 1 year (25/qtr)
Forearms	30 rems in any 1 year (25/qtr)
Other organs, tissues, and organ systems	15 rems in any 1 year
Fertile women (with respect to fetus)	0.5 rem in gestation period
Dose limits for the public or occasionally exposed individuals	0.5 rem in any 1 year
Students	0.1 rem in any 1 year
Genetic cells in total population	0.17 rems average per year
Somatic cells in total population	0.17 rems average per year
Emergency doses > 45 years old	100 rems
Emergency doses, hands and forearms > 45 years old	200 rems additional to above for a total of 300 rems
Emergency dose limits—less urgent	
Individual	25 rems
Hands and forearms	100 rems
Family of radioactive patients	
Individual < 45 years old	0.5 rems in any 1 year
Individual > 45 years old	5 rems in any 1 year

OSHA exposure levels for confined spaces is given in table G-18, 29 CFR Part 1910.96. The table is modified below.

Table 9-5. Modified version of OSHA's permissible levels of radiation exposure in restricted areas.

Exposure	Rems per calendar quarter
Whole body: head and trunk; active blood-forming organs; lens of eye; or gonads.	1.25 rems per quarter
Hands and forearms; feet and ankles.	18.75 rems per quarter
Skin of whole body.	7.5 rems per quarter

Table 9-6. DOT-referenced regulations for the shipment of NORM.

Transportation Application	Code of Federal Regulations
Shipping papers	49 CFR 172.201-204
Markings	49 CFR 172.301 and 310
Labeling	49 CFR 172.400, 403, and 406
Packaging	49 CFR 173.15 and 425 (a) and 178.350
Placarding	49 CFR 172.504
Emergency response information	49 CFR 172.600
Hazardous materials revisions	Federal Register December 21, 1992 (55 FR 52402)—expected to be promulgated by October, 1993.

PROTECTION DURING THE TRANSPORTATION OF NORM

Transportation of NORM is controlled by the U.S. Department of Transportation (DOT). Table 9–6 outlines the requirements and reference to the applicable Code of Federal Regulations. DOT regulations define oil and gas field NORM wastes, as material containing 2,000 pCi/gram to 100,000 pCi/gram. DOT has classified these materials as Low Specific Activity (LSA). Note the lack of classification below 2,000 pCi/gram.

CHAPTER 9
REFERENCES

1. McGuire, S.A., and Peabody, C.A., *Working Safely with Gamma Radiography*, Office of Nuclear Research, U.S. Nuclear Regulatory Commission, 1982.

2. Bollinger, N.J., and Schutz, R.H., "NIOSH Guide to Industrial Respiratory Protection," U.S. Department of Health and Human Services, 1987.

3. International Atomic Energy Commission, Vienna, *Safe Handling of Radioactive Nuclides*, Safety Series No. 1.

4. Plog, B.A., Benjamin, G.S., and Kerwin, M.A., *Fundamentals of Industrial Hygiene*, Third Edition, National Safety Council, 1990.

RADIATION INSTRUMENTATION

INTRODUCTION

Radiation measurement can be accomplished with several different tools. The most commonly used tool in the past was the Geiger-Muller tube. Scintillation counters are now employed and generally more sensitive than Geiger-Muller tubes. Whatever instrument is put to use to detect NORM, it must be calibrated periodically by the manufacturer or a certified laboratory. A third, less sensitive instrument, is the dosimeter, which must be charged after each survey. This instrument is often used in situations where equipment weight or size might present a problem, as it will fit in a shirt pocket. The dosimeter has a sensitivity of about 10 %+ − and is a very rugged, compact instrument. Other types of personal monitoring instruments include film badges, radon detectors, and thermoluminescence detectors.

GEIGER-MULLER TUBES

A Geiger-Muller counter is used for gamma, beta, and X ray detection. It is especially sensitive to *beta radiation*. It utilizes an ionization tube that is filled with a special gas. The tube is fitted with a thin glass or plastic window through which radiation enters. Argon gas is often used to fill the tube. A wire runs down the center and is insulated from the tube. The tube and the wire are connected to a high-voltage source, and the tube becomes the negative electrode and the wire becomes the positive electrode. The gas in the tube serves as an insulator until a radioactive particle or gamma ray passes through the gas. The atoms of gas are then ionized. A very small number of ions are needed to place the tube into a discharge condition. Electrons are freed by the initial ionization process, and they acquire enough extra energy by the applied voltage to create additional ions. The central wire must be very small to produce an avalanche of ionization. The wire is typically less than 0.01 inches in diameter. After the counter discharges, there is a dead time between counts, which may last approximately 100-200 microseconds. This dead time may miss a few counts. A large number of quenching vapors may be utilized within the glass tube. Cleanliness is an absolute factor for the interior of the glass. A typical counter is filled with ethyl alcohol to a pressure of 1 cm of Hg (mercury) and argon to a pressure of 9 cm of Hg, operating at a voltage of 1,000 volts. The glass tube is typically 1 inch in diameter. The

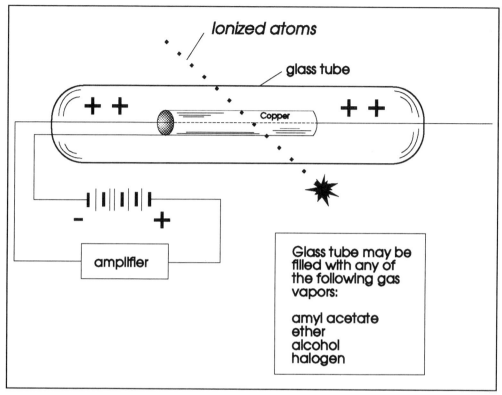

Figure 10-1. A Geiger-Muller radiation counter.

output pulse is relatively constant. The high field gradient in the immediate vicinity of the wire causes electrons (negative charge) to accumulate toward the central copper electrode and attain high kinetic energies. The Geiger-Muller counter does not give a uniform response for different types of radiation. It is accurate for only the type of radiation energy for which it has been calibrated to measure.

SCINTILLATION COUNTER

A scintillation counter has the advantage of not having to utilize an enclosed ionization tube, which may miss an alpha particle. The scintillation counter utilizes a

Figure 10-2a-b. A scintillation counter showing the two type of probes. Ludlum utilizes a Geiger-Muller probe for alpha and beta radiation attached to the same amplifier and counting device. Photos provided through the courtesy of Ludlum Measurements, Inc., Sweetwater, Texas.

thallium-doped sodium iodide crystal (NaI [Ti]), as the prime detection sensor. The detector responds to radiation by emitting a visible flash of light. The flashes of light are then detected by a photomultiplier. Plastic (organic) scintillators are commonly used for counting beta radiation and alpha particles. Another good detecting crystal is zinc sulfide.

Probes have been designed to detect alpha, beta, and gamma radiation. The probes that measure alpha and beta radiation place the crystal in an uncased position, covered only by a screen or thin piece of muscovite or Mylar. The size and shape of the crystal determines its sensitivity. The larger crystals are expensive and difficult to grow, but provide a very high degree of sensitivity. The scintillation counter is far more versatile than the Geiger-Muller counter when working with NORM.

DOSIMETER (ELECTROSCOPE)

The dosimeter consists of an ion chamber and a quartz fiber volt meter. It is similar to a Geiger-Mueller counter in that it is sensitive to gamma and beta radiation, but is limited in the type and accuracy of radiation it can detect. An insulated fiber is supplied with a static charge relative to another electrode. The electrostatic force placed on the fiber causes it to deflect. The deflection corresponds to a fixed referenced voltage, which is determined to be a zero dose of radiation on a scale. As the electroscope (dosimeter) is exposed to radiation, the referenced charge leaks away through a gas-filled tube, causing the fiber to deflect less from the referenced voltage. This lessening deflection is calibrated to a scale which corresponds to terms of total radiation dose. The applications are confined to time measurements of hours or days.

Various chemical dosimeters have been manufactured based upon the ability of ionizing radiation to induce chemical reactions. Some of the chemical

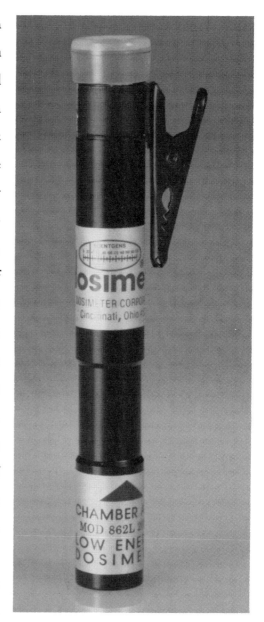

Figure 10-3. A dosimeter radiation counter. Photo provided through the courtesy of the Dosimeter Company.

dosimeters utilize ferrous-ferric chemicals or cerous-ceric chemicals. The latter is used for very high rates of radiation exposure. The ferrous-ferric type of dosimeters involve the conversion of ferrous ions to ferric ions in a dilute solution of sulfuric acid (0.8%) through the process of radiation induced oxidation.

Dosimeters have been developed to alter the color of dyes as radiation is absorbed into the dye material. One of the elements that has been used in this type of dosimeter, in the form of a metal foil, is dysprosium, a rare earth element. This type is perhaps the least accurate.

Radon Detectors

Several types of relatively inexpensive radon detectors have been developed for household and industrial use. They range from sophisticated electronic instruments, which digitally accumulate counts and print them out on a thermal printer, to simple film cartridges. Some of the digital models may just display the count without all the other bells and whistles. The costs range from approximately $50.00 to well over $2,000.00. Many real estate management companies now require radon measurements for residential sales inspection, prior to closing. It is commonly performed in the most radon-prone parts of the country.

Chapter 10
References

1. Ludlum Measurements, Inc., *Instruction Manual, Model 19, Micro R Meter,* 1992.

CHAPTER 11

SURVEYS AND SAMPLING

INTRODUCTION

There are essentially two types of field surveys in the NORM arena. They are *equipment* surveys and *geological* surveys. The equipment surveys may include any manufactured item, such as a valve, a piece of tubing, a pump, a heater treater, coal ash scrubber, bag filter, etc. Geological surveys might include contaminated land, household radon, ground water, coal, lignite, methane, crude oil, phosphate waste, or any produced or any mined substance that contains NORM. It is all naturally occurring, but may be concentrated through processing and turbulent flow of water, petroleum, or gas.

There are two types of measurements. First, there are *field measurements* and second, there are *laboratory analyses*. Field measurements require specific instruments and measuring techniques. This chapter is devoted to the surveying and sampling techniques applied to NORM. The reader is encouraged to keep up with the rapidly evolving changes in the field of sampling and survey techniques.

FIELD SURVEYS

Equipment radiation can either be found in scrapped or working equipment. It is often necessary to change out pump gaskets or replace valve components. Whenever equipment is opened for repair or discarded in a NORM setting, it should be surveyed. NORM accumulates in equipment by *turbulent flow* (see Figs. 11-1, 11-2). It builds up as scale or sludge on the internal components of equipment, especially at constriction points or changes in direction points. The valve depicted in Figure 11-2 shows the areas of NORM buildup in restricted flow areas. *Gamma* radiation may be present *externally*, thus, prior to opening any valve or device for repair or discard, it should be surveyed from the *outside first*. If gamma radiation is detected from the outside of the device, it is certain that safety measures must be taken to handle the equipment. It is good practice to survey all constricted-flow equipment on a quarterly basis by trained personnel and to keep a statistical record of the surveys. The changes should be noted through time. It is not practical, of course, to apply quarterly surveys to down-hole tubing, but the surface equipment should be checked on a quarterly basis. This is especially true for oil and gas wells as they begin to produce brine

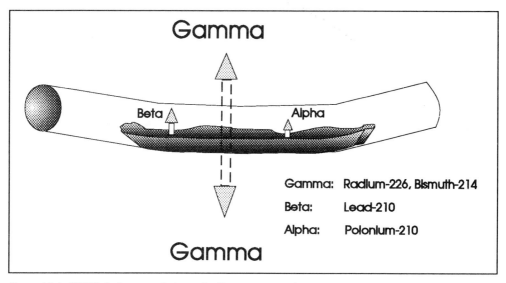

Figure 11-1. NORM sludge or scale in production process equipment.

along with the petroleum. Field personnel can be trained to perform these checks and should be instructed to maintain accurate records. When tubing is pulled during workover operations, it should be surveyed as it comes out of the hole to ensure that adequate safety measures are adhered to, if the situation merits NORM concern. The American Petroleum Institute (API)[1] has published a map of areas in the U.S. that may contain NORM. There will be many other areas that will not receive the research and scrutiny that has just begun to unfold. NORM will undoubtedly be discovered in other portions of the United States and the world as awareness spreads. The operator should not jeopardize the employees who handle NORM, only to find out 30 years later that they have lost eye lenses or developed cancer from handling an unknown quantity of NORM earlier in their careers.

DETECTING RADIATION IN SCRAP METAL

The detection of radiation in scrap metal is looming as an issue of greater concern. For steel mills, the potential exists for multi-million dollar lawsuits, substantial cleanup costs, worker injury, and consumer concern. For scrap metal suppliers, the expense of rejected loads is becoming higher. Field measurements of scrap loads are conducted by scintillation counters mounted on both sides of a check point, as trucks are permitted into the scrap metal yard.[2] Figure

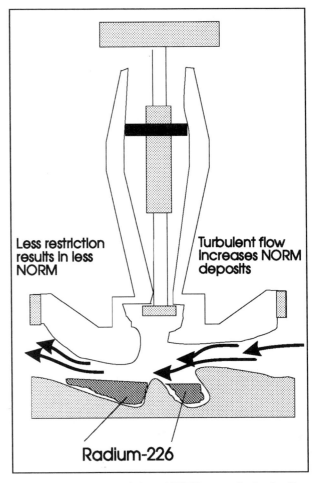

Figure 11-2. The accumulation of NORM contamination is often increased when the flow direction of the product is altered or restricted during processing.

11-3 displays the basic setup at a scrap metal check point.

Fluctuations in background counts can set off alarm systems, thus requiring a site survey by hand. Trucks filled with scrap must stop at these automatic check points prior to their destination within the scrap yard. This may also be applied in the future to weigh stations along the major highways to comply with DOT shipping

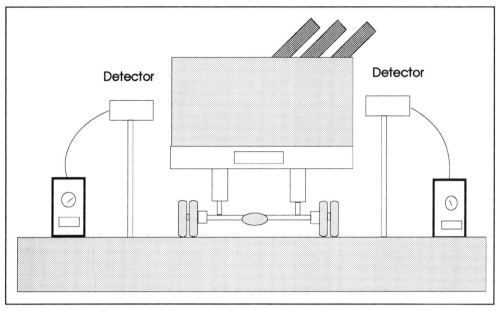

Figure 11-3. A truck passing through a scrap metal radiation check point.

and labeling regulations. A good system will cost about $90,000.00, and is based upon a microprocessor alarm system that must be frequently calibrated. Many scrap dealers have now installed radiation detection equipment. This is especially true of the dealers that receive metal from the petroleum production, transportation, and petrochemical industries. If a load of scrap is rejected, there are only three things an operator can presently do at that point:

1. Store the contaminated scrap in properly marked drums on their own premises. In the case of leased property, it would behoove the landowner to be aware of this option.

2. Cut up the metal and place it with concrete in properly marked drums and ship it to the only facility in the country, located in Utah.

3. With the proper permits, grind it up into fine powder and inject it into a hazardous waste well (Class I) and/or a Class II brine well. There are presently about

1,200 Class I wells and about 170,000 Class II wells. It is anticipated that new federal restrictions will soon be imposed on Class II injection wells. They may include associated NORM regulations.

As awareness of the problem of NORM becomes more widely known, new solutions for disposal will be forthcoming, but not without great controversy and potential expense. It is the opinion of the author that the two most viable options for NORM disposal are careful, geologically selected, landfill areas and prudently regulated Class II injection wells.

Scale Removal and Disposal

Scale can be removed from contaminated equipment by sandblasting, water jetting, or chemical removal. It is more cost effective to use a mechanical means than chemical. Worker exposure becomes an acute safety problem if precautionary measures are not taken. The disposal of waste then becomes the next issue. Prevention may be a future alternate solution, but will require new technology and careful monitoring.[3]

Origin and Nature of NORM in Petroleum Production Equipment

The NORM that accumulates in surface petroleum production equipment is predominantly Ra^{226} and Ra^{228} and their daughter elements. Radium decay products are slightly soluble and under some circumstances become mobilized by liquid phases in their formation and deposition. When brought to the surface with liquid production streams, these radionuclides may remain dissolved at dilute levels or may precipitate, due to chemical changes at reduced pressures and temperatures, as the fluids are separated from particulate matter in the flow process. Since radium concentrations in the original geological formations are highly variable, production fluids are also highly variable. Fluids injected into the formations also affect the mobilization of NORM, and surface processes further vary the accumulation of any radioactive scales, sludges, and/or wastes.

The NORM accumulated in production equipment typically contains radium co-precipitated in barium sulfate ($BaSO_4$). Sludges are dominated by silicates or carbonates, but also incorporate varying amounts of radium by co-precipitation. NORM deposits may accumulate in gas plant equipment from Rn^{222} (radon gas). The more mobile radon gas primarily originates in subsurface geologic units and becomes dissolved in brines and in the organic petroleum fractions in gas plants. When NORM substances enter the surface equipment it is partitioned mainly into the propane and ethane fractions because of its solubility. Gas plant deposits differ from oil production scales and sludges by having very low mass. NORM typically consists of an invisible film of a radon daughter, Pb^{210}, which accumulates on the interior surfaces of pipes, valves, pumps, and other associated equipment. Oil and gas field audits should always include radiation sampling of all surface equipment, including gas plants.

Petroleum industry wastes in equipment can be classified into four categories, based upon different characteristics that stem from NORM. These categories include the following list.

1. Sludges found in petroleum production equipment.

2. Scales found in tubing, pipes, and other production equipment.

3. Equipment that contains residual NORM scale left after cleaning.

4. Thin deposits of radioactive lead, Pb^{210}, found on the interior of pipes,

valves, pumps, and other equipment associated primarily with gas plants and compressor stations.

The characteristics of each NORM group are described in this chapter to define the basis for estimating radiation exposure. Since waste characteristics affect exposures by different pathways, a characteristic may be more important in some disposal alternatives than in others.

SLUDGES

Sludges that accumulate in production equipment typically contain Ra^{226} and Ra^{228} concentrations ranging from background levels to several hundred pCi/gram. Radium^{226} concentrations dominate the NORM-contaminated sludge. Since both radium decay chains exhibit similar types of radiation, they are generally expressed in total radium. Radium^{226} will make up about 78% of the total radium, while Ra^{228} comprises approximately 22%. Sludges typically have a granular consistency, dominated by a bulk composition of silicates and carbonates. Bulk densities in equipment or disposed deposits are typically about 1.6 g/m^3, with porosities around 39%. Radium in sludge has a distribution coefficient for solid/aqueous phases of approximately 2,500 cm^3/g. Lead, Pb^{210}, has a distribution coefficient of about 5,100-20,000 cm^3/g. The distribution coefficient helps define the leach characteristics for ground water exposure pathways.[2]

SCALES

Scales accumulate in tubing, separators, and other equipment. They contain a broader range of Ra^{226} and Ra^{228} concentrations, ranging from background levels to several thousand pCi/gram. Radium^{226} again makes up the bulk of the total radium present as NORM scale. The scales occur as very hard, monolithic precipitates in equipment, with bulk densities that range from 2-3 g/cm^3. If they are removed in a disposal process, their bulk densities fall to around 1.6 g/cm^3, due to the very high porosities that average about 45%. Radium in scales has a distribution coefficient for the solid/aqueous phases of 250,000 cm^3/g, which defines once again the leach characteristics for ground water pathways.

PRODUCTION EQUIPMENT

Residual NORM remaining in production equipment after cleaning usually occurs in scales, since they are tightly attached to equipment surfaces and are largely insoluble. Typical scale thickness varies from less than 0.1 in. to several inches in production tubing, mainly in brine water lines. Total radium concentrations and radon emanation fractions compose the bulk of the radioactive material in the scales. Densities of disposed equipment vary due to equipment geometry. However, porosities are typically large and NORM sites of disposed equipment are small, due to dilution of the equipment mass. The volume of scales remaining in equipment, if no mechanical cleaning is done prior to disposal, ranges from about 1% to 77%, with an average of about 6.7% of the total equipment volume. Leaching characteristics are similar to those listed under the sub-heading for NORM scale.

GAS PLANT EQUIPMENT

Thin layers of Pb^{210} deposited from radon daughters form on the inside surfaces of gas plant equipment and differ from other NORM accumulations. They have negligible mass, being invisible and containing only the last three radionuclides from the U^{238} decay series. Activity concentrations are expressed in units or radioactivity per unit area of equipment, since the mass of the NORM is not measurable or of interest. Occurrences range from background levels to several thousand disintegrations per minute in a 100 cm^2 area (dpm/100 cm^2). No gaseous radon effluents are associated with these deposits. Leaching characteristics are dominated by the leachability of lead and polonium. Since they occur primarily in gas plant equipment and compressor stations, they are usually associated with massive equipment upon disposal. Abrasive cleaning of metal surfaces with coatings of NORM that approximate 0.004 inches in thickness is assumed to be removed. This should always be checked prior to disposal or placement into service.

RADIATION CONTROL AREA

Radiation contamination should be treated in a similar fashion to a hazardous waste site, when working around a high concentration of NORM. Project assessments should be carefully reviewed to ensure that the designated control area be checked, both before entering and exiting, and that it provides an adequate work space. The following guidelines should be followed:

1. Only D.O.T. Type 17H 55-gal. drums or equivalent should be implemented for control, storage, transportation, or disposal of NORM.

2. Control area perimeters should be enforced with ribbon and signs to keep unauthorized personnel out of the control area.

3. Only one entrance and exit area should be designated with an adjacent decontamination area. Workers should be checked for radiation upon both entering and exiting. Records should be kept on each radiation check.

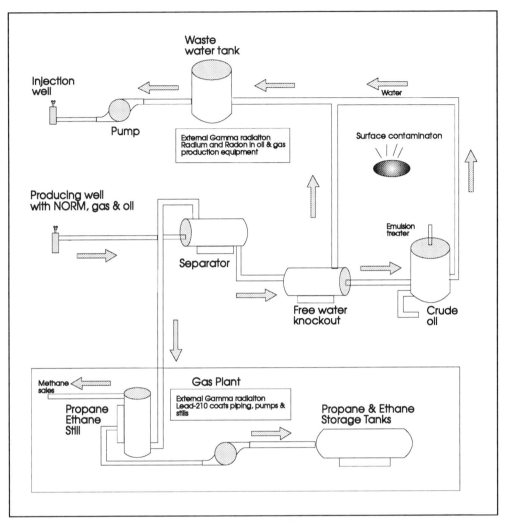

Figure 11-4. Gas plant layout with NORM types and points of concentration.

TECHNIQUES OF GROUND NORM RADIATION SURVEYS

Geological surveys are affected by radiation emanating from the surface and sub-surface. Topography can have a very large effect on the individual stations that are

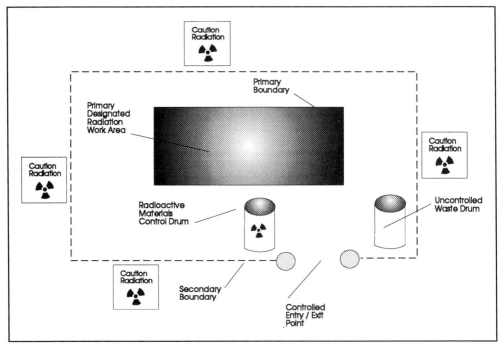

Figure 11-5. Elements of a control radiation work area.

metered. Generally a grid is laid out on some appropriate spacing, depending upon the area being surveyed and the detail the survey might require. The geologist must make a judgmental decision on the size of the grid for each site. Grids may vary from 10 meters to several centimeters. A grid might be laid out on a 1 X 1 m grid or a 10 x 10 m grid, or some other appropriate scale. Random sampling may be an option at some sites. This would consist of assigning numbers to each grid square and then statistically generating a range of numbers within that range laid out on the grid and sampling those randomly generated grid points. A more precise method might be to survey entire profiles at some specified interval. As radiation "hot spots" are discovered, then systematically concentrate on those areas. Figure 11-6 depicts two methods of sampling a grid, which include

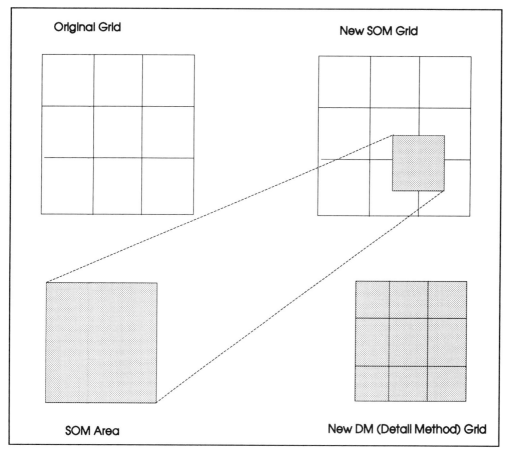

Figure 11-6. One grid method and a detail method.

the Statistical Offset Grid (SOG) and the Original Grid (OG).

The Louisiana Department of Environmental Quality (LDEQ) recommends specific sampling procedures for several geometric types of contamination.[4] The LDEQ requires the sampling program to be comprehensive, covering the entire area. They state "...licensee's survey should be on grid spacing small enough to assure the division, that a contaminated area has not been missed..." They further state that if 50% of the area is contaminated, then 50% of the samples should come from the contaminated

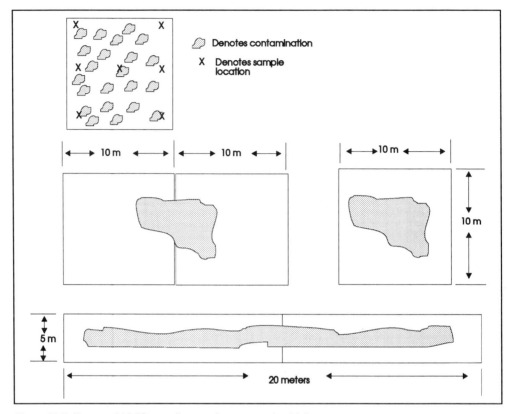

Figure 11-7. Illustrated LDEQ sampling requirements and guidelines.

area. A minimum of five samples shall be taken from a 100 m^2 area to a minimum depth of 15 cm. If a small area of contamination is found, then sampling should place the contaminated area within the center of a 100 m^2 grid. They require sampling to be statistically defensible. Figure 11-7 shows the various LDEQ requirements.

STATISTICAL DETERMINATION OF GRID SPACING

The following statistical procedure might be employed to maintain statistical viability.[5] It is assumed that a contaminated area has been defined.

1. Specify L, the length of the semi-major axis of the smallest hot spot important to detect. L is 1/2 the length of the long axis of the ellipse.

2. Specify the expected shape (S) of the elliptical target, where:

$$S = \frac{\text{length of short axis of the ellipse}}{\textit{length of long axis of the ellipse}}$$

Note that $0 < S \leq 1$ and that $S = 1$ for a circle. If S is not known, a conservative approach is to assume a skinny elliptical shape perhaps on the order of $S = 0.5$

3. Specify an acceptable probability (β) of not finding a hot spot. The value of β is known as the *consumers' risk*. For example, we might be willing to accept a 100 X β% = 20% chance of not finding a small hot spot of contaminated material.

4. Refer to Figures 11-8, 11-9, and 11-10 for square, rectangular, and triangular grids. These nomographs offer the relationship between β and the ratio of L/G and G is the spacing between grid lines. Using the curve corresponding to the shape (S) of interest, find L/G on the horizontal axis that correlates to the pre-specified β. The total number of grid points (sampling locations) can then be found. The dimensions of the sample area are known.

EXAMPLE:[5] Assume a square grid is used, and it is desired to take no more than 100β = 10% chance of not hitting a circular target of radius L = 100 cm or larger. Utilizing the curve depicted in Figure 11-8 for $S = 1$, it is found that L/G = 0.56. Thus, 100 cm/0.56 $\approx$ 180 cm. Hence, if cores are taken on a square grid with spacing of 180 cm, it can be assured the probability is only 0.10 (1 chance in 10) of not hitting a circular target that is 100 cm or more in radius.

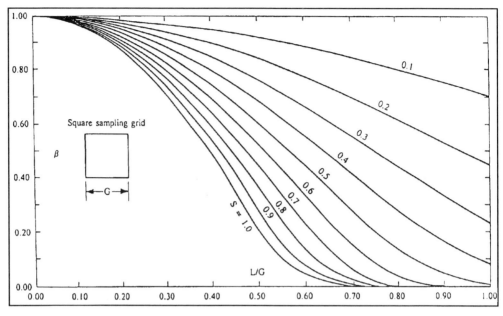

Figure 11-8. Curves relating to L/G to consumers' risk, β, for different target shapes, when sampling is on a square grid pattern.[5] Printed by permission from Van Nostrand Reinhold, *Statistical Methods for Environmental Pollution Monitoring*, by Richard O. Gilbert.

SURVEY RESPONSE TO GEOLOGIC AND TOPOGRAPHIC ARCHITECTURE

Geology and topography will affect specific survey readings, and the geologist must account for these readings when constructing contour maps of radiation levels. If these factors are not taken into consideration, an erroneous conclusion might be reached. The reader is referred to Figure 11-11 (pg. 181), which is accompanied by explanations (pg. 179). The reader should also be aware of the dampening effect of radiation in water saturated soils, swamps, and wetlands.

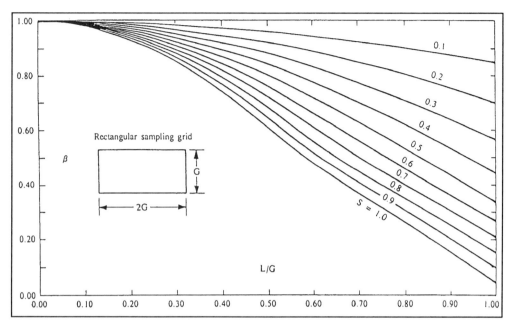

Figure 11-9. Curves relating L/G to consumers' risk, β, for different target shapes, when sampling is on a rectangular grid pattern.[5] Printed by permission from Van Nostrand Reinhold, *Statistical Methods for Environmental Pollution Monitoring*, by Richard O. Gilbert.

A. This scene depicts a ground level (solid line) reading and a 1-m reading (dashed line) with a scintillometer. Note the higher reading above the ground level, indicating a mass effect from a larger area. The contaminated zone is labeled.

B. This scene depicts a reading about $^{1}/_{2}$ m into the first zone, which is closer to the radioactive zone.

C. This scene displays the scintillation probe very near the top of the radioactive zone. Note the higher reading.

D. This scene displays the probe below the radioactive zone. Note the lower reading.

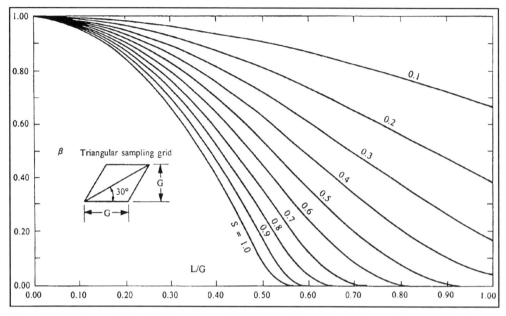

Figure 11-10. Curves relating L/G to consumers' risk, β, for different target shapes when sampling is on a triangular grid pattern.[5] Printed by permission from Van Nostrand Reinhold, *Statistical Methods for Environmental Pollution Monitoring*, by Richard O. Gilbert.

CHAPTER 11
REFERENCES

1. American Petroleum Institute, Bul. E2: *Management of Naturally Occurring Radioactive Materials (NORM) in Oil and Gas Production*, 1992.

2. Suntrac Services, Inc., Naturally Occurring Radioactive Seminar, University of Houston Institute for Environmental Management, 1992.

3. Connor, J., Clifford, D., and King, P.T., "In-Situ Method for Reducing the Production of Naturally Occurring Radioactive Material (NORM) from Subsurface Reservoirs," Book 3, p.461, Petro-Safe Conference, 1993.

4. Louisiana Department of Environmental Quality, Radiation Protection Division,

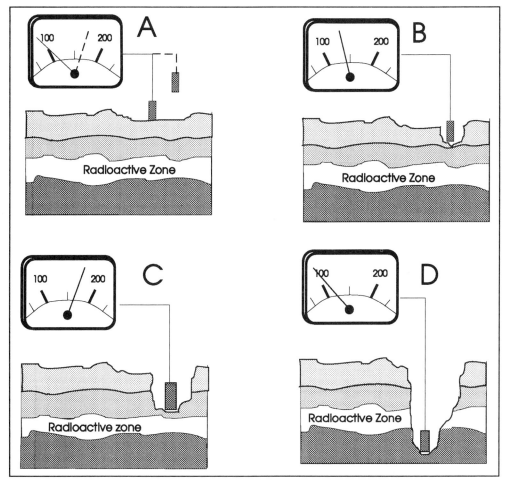

Figure 11-11. Four scenes demonstrating the difference in scintillator readings from differing levels of topography. It may be viewed as a ditch on a local level by changing scale.

Implementation Manual for Management of NORM in Louisiana, 1992.

5. Gilbert, R. O., *Statistical Methods for Environmental Pollution Monitoring*, Van Nostrand Reinhold, New York, 320 pages, 1987.

BIBLIOGRAPHY

NORM LITERATURE

Abelson, P.H., "Uncertainties about Health Effects of Radon," *Earth Science*, v. 250, p.353, 1990.

Adams, J.A.S., *The Natural Radiation Environment*, U. Of Chicago Press, 1964.

AIPG, "Radioactive Waste, Issues and Answers," American Institute of Professional Geologists, August, 1985.

AIPG, *Hazards of Geologic Materials*, 1993.

American Metals Market, "ISRI Targets Radioactive Scrap," p.8, February 19, 1992.

API: Comm 1988, *NORM Regulatory Analysis Report*, Ad Hoc Committee on Naturally Occurring Radioactive Material, American Petroleum Institute, April 1988.

API: Miche, T., *Oil and Gas Industry Water Injection Well Corrosion*, Feb., 1987.

API: Otto, G.H., *A National Survey on Radioactive Material (NORM) in Petroleum Producing and Gas Processing Facilities*, American Petroleum Institute, 1973.

API: Rogers & Associates, Inc., *Methods for Measuring Naturally Occurring Radioactive Materials in Petroleum Production Equipment*, American Petroleum Institute, 1990.

API: Snavely, E.S. Jr., *Radionuclides in Produced Water*, American Petroleum Institute, 1989.

API Bul E2: *Management of Naturally Occurring Radioactive Materials (NORM) in Oil & Gas Production*, American Petroleum Institute, 1992.

Bohlinger, L.A., "Regulation of Naturally Occurring Radioactive Materials in Louisiana," 22nd Annual Meeting, Conference of Radiation Control Program Directors, Salt Lake City, Utah, 1990.

Brookings, D.G., *The Indoor Radon Problem*, Columbia UniversityPress, 1990.

Brown, D.A., *Boston College Environmental Affairs Law Review*, Vol. 16, No. 2, Winter, 1988.

Burr, B.J., Howe, T.M., and Goulding, J., "The Development and Application of a Detectable Polymer Scale Inhibitor to Control Sulfate Scales by Squeeze Applications," SPE, International Symposium on Oil field Chemistry, San Antonio, Texas, 1987.

Cayias, J.L., Gordon, R.D., Davidson, K.B., "NORM Waste in Oil and Gas Operations: Management, Regulations and Disposal," Petro-Safe '93 Conference Proceedings, January, 1993.

Cohen, B., *Radon—A Homeowner's Guide to Detection and Control*, New York, Avon Books, 1989.

Connor, J.A., Clifford, D., and King, P.T., "In-Situ Method for Reducing the Production of Naturally Occurring Radioactive Materials (NORM) From Subsurface Reservoirs," Petro-Safe '93 Conference Proceedings, January, 1993.

Crawford-Brown, D.J., *Radon, Radium, and Uranium in Drinking Water*, Lewis Publishers, Inc., Chelsea, MI, 1990.

CRCPD, *Draft 6, Rationale Part N*, "Suggested Regulations for Control of Radiation" (Updated), Council of Radiation Control Program Directors, Frankfort, Kentucky.

CRPCD, "Part N Regulation and Licensing of Naturally Occurring Radioactive
    Materials (NORM)," Council of Radiation Control Program Directors,
    Frankfort, Kentucky, November, 1990.

Darnel, T., "Radioactivity in My Backyard?" *American City and County*, August,
    1990.

"Detecting Radioactive Materials in Scrap," *Modern Casting*, p. 29, September, 1991.

Drummond, I., et al. "Occurrence of Rn^{222} and Progeny in Natural Gas Processing
    Plants in Western Canada," *Health Physics*, November, 1990.

Duval, J.S., Riggle, F.B., Jones, W.J., and Pitkin, J.A., "Equivalent Uranium Map
    of the Coterminous United States," U.S.G.S. Open File Report
    89-478, 1989.

Egan, J.R., and Seymour, J.F., "Disposing of Naturally Occurring Radioactive
    Material Wastes," *Environmental Law Reporter*, Vol. 22, No. 7, p.10433(7)
    July, 1992.

EPA Bandrowski, M.S., "Regulation of Higher Activity NARM Waste by EPA,"
    DOE's 10th Annual LLW Management Conference, Denver Colorado,
    September, 1988.

EPA: *Risk Assessment Methodology: Environmental Impact Statement for National
    Emission Standards for Hazardous Air Pollutants*, Vol. 1, Background
    Information Document, Office of Radiation Programs, U.S. Environmental
    Protection Agency, September, 1989.

EPA: "Suggested Guidelines for the Disposal of Drinking Water Treatment Waste
    Containing Naturally Occurring Radionuclides," Office of Drinking Water,
    U.S. Environmental Protection Agency, July, 1990.

EPA: "National Primary Drinking Water Regulations: Radionuclides," Federal
    Register, Vol. 56, No. 138, July 16, 1991.

EPA: *Radon Reduction Methods*, RD-681, U.S. Environmental Protection Agency,
    July, 1989.

EPA, *National Radon Measurement Proficiency (RMP) Program*, Texas State
    Proficiency Report, U.S. Environmental Protection Agency,
    EPA 52Q/1-91-014, December, 1991.

EPA: *Removal of Radon From Household Water*, U.S. Environmental
    Protection Agency, OPA-87-001, September, 1987.

EPA: *A Citizens Guide to Radon*, 2nd Edition, U.S. Environmental
    Protection Agency, 402-K92-001, May, 1992.

"EPA Rules on Mineral Processing Wastes," New Jersey Industry Environmental
    Alert, July, 1991.

EPA / SCA: "Diffuse NORM Waste: Waste Characterization and Risk
    Assessment (Draft)," SC and A Inc., for Office of Radiation Programs, U.S.
    Environmental Protection Agency, May, 1991.

Gates, A.E. and Gundersen, L.C., "Geologic Controls on Radon," Geol. Soc. of
    Am. Special Paper in press.

Gessell, T.F., "Occupational Radiation Exposure Due to Rn^{222} in Natural Gas and
    Natural Gas Products," Health Physics, Vol. 29, November, 1979.

Gilges, K., "Europe's Hazardous Waste Dilemma," Chemical Engineering, Vol. 98,
    No. 8, p.30, August, 1991.

Gott, G.B., and Hill, J.W., "Radioactivity in Some Fields of Southeastern Kansas,"

U.S.G.S. Bulletin 988-E, U.S. Government Printing Office, Washington, D.C., 1953.

Graves, B., *Radon in Groundwater*, Chelsea, MI, Lewis Publishers, 1987, available through the National Ground Water Association.

Gray, P.R., "Radioactive Materials Could Pose Problems for the Gas Industry," *Oil & Gas Journal*, June 25, 1990.

Gray, P.R., "Regulations for the Control of NORM," SPE 26272, EPA/SPE Environmental Conference Exploration and Production, San Antonio, March, 1993.

"The Growing Puzzle Over Hot Scrap," *Recycling Today*, Vol. 29, N. 8. p. 68, August, 1991.

Gundersen, L.C.S. and Wanty, R.B., "Field Studies of Radon in Rocks, Soils, and Water," U.S.G.S. Bulletin 1971, 1991.

Hallenbeck, W.H., and Cunnigham, K.M., *Quantitative Risk Assessment for Environmental and Occupational Health*, Lewis Publishers, Inc., 1988.

"Hearing on Nuclear Waste Sites to Begin June 27," Latin American Business News Wire, June 4, 1992.

Heyer, "The Status of Texas' Regulatory Effort Concerning Naturally Occurring Radioactive Material," Proceedings: Eleventh Annual Gulf of Mexico Information Transfer Meeting, New Orleans, Louisiana, November, 1990.

Horton, T.R., *Nationwide Occurrence of Radon and Other Natural Radioactivity in Public Water Supplies*, EPA, 520/5-85-008, 1985.

Implementation Manual for Management of NORM in Louisiana, Louisiana Department of Environmental Quality, June, 1992.

Johnson, R.H, Jr., *Assessment of Potential Radiological Health Effects from Radon in*

Natural Gas, U.S. Environmental Protection Agency Report, EPA-520/1-73-004, 1973.

Kramer, T.F., and Reid, D.F., "The Occurrence and Behavior of Radium in Saline Formation Water of the U.S. Gulf Coast," *Isotope Geoscience*, Elsevier Publishing, Vol. 2, 1984.

Landa, E.R, and Reid, D., "Sorption of Radium[226] From Oil Production Brine," *Environmental Geology*, Vol. 5, No. 1, 1983.

Lowe, D.J, "NORM, Cleaning and Disposal Using Closed Loop Hydro-Blasting of Solvent Bath and Underground Injection System," EPA/SPE Environmental Conference Exploration and Production, San Antonio, Texas, March, 1993.

MacNamee, "Cleaning of Naturally Occurring Radioactive Materials and North Sea Procedures," Proceedings: Eleventh Annual Gulf of Mexico Information Transfer Meeting, New Orleans, Louisiana, November, 1991.

Marshall, A., "Radon Found at Gas Plants," *Calgary Herald*, Letter to the Editor, from Martha Kostuch, September 24, 1992.

May, C., "Radiation From Radium in Oil Fields Could Prove Costly for Industry," *The Energy Report*, Vol. 18, p.778, December 17, 1990.

McGuire, S.A. and Peabody, C.A, "Working Safely With Radiography," Office of National Council on Radiation Protection and Measurements, *Evaluation of Occupational and Environmental Exposures to Radon Daughters in the United States*, NCRP, Report 78, Bethesda, MD, 1984.

"Mississippi Suits Could Decide Oil Companies' Liability for Cleanup of Radioactive Wastes," *Nuclear Waste News*, Ziff Communications Co., January 10, 1991.

"The New Hot Wastes," *Science News*, p.264-167, October 26, 1991.

NRC / Donnell, E. and Lambert, J., "Low Level Radioactive Waste Research
	Program Plan," Division of Engineering, Nuclear Regulatory Research, U.S.,
	Washington, D.C., 1989.

Nuclear Regulatory Research, U.S. Nuclear Regulatory Commission,
	NUREG/BR-0024, September, 1982.

Oddo, J.E., and Sitz, C.D., "NORM Scale Formation: A Case Study of the
	Chemistry Prediction, Remediation, and Treatment in Wells of the Antrim
	Gas Field," EPA/SPE Environmental Conference Exploration and
	Production, San Antonio, Texas, March, 1993.

OTA, "Partnership Under Pressure: Managing Low Level Radioactive Waste,"
	Office of Technology Assessment Congress of the United States, 1989.

Otton, J.K., *The Geology of Radon*, U.S.G.S. Publication, 1992.

Pierce, A.P., et al., "Uranium and Helium in the Panhandle Gas Field, Texas and
	Adjacent Areas," U.S.G.S., Professional Paper 454-G, U.S. Government
	Printing Office, Washington, D.C., 1964.

"Pipeliners Must Deal with New Environmental Challenges," *Pipeline and Gas
	Journal*, p.53, February, 1992.

Renfro, J.J., "NORM, Naturally Occurring Radioactive Materials in
	the Oil and Gas Industry," *Training Manual*, December, 1990.

Rogers & Associates, "Safety Analysis for the Disposal of Naturally Occurring
	Radioactive Materials in Texas," Texas Low Level Radioactive Waste
	Disposal Authority, Austin, Texas, October, 1988.

Rutherford, G.J, and Richardson, G.E., "Disposal of Naturally Occurring

Radioactive Materials," EPA/SPE Environmental Conference Exploration and Production, San Antonio, Texas, March, 1993.

Schneider, K., "Radiation Dangers Found in Oil Fields Across the Nation," *New York Times*, December 3, 1990.

Schuman, R. R., and Gundersen, L.C.S., and Tanner, A.B., "Geology and Occurrence of Radon," *Measurement, Prevalence, and Control*, ASTM MNL 15, 1993.

Shannon, B.E., "An Operational Perspective on the Handling and Disposal of NORM in the Gulf of Mexico," EPA/SPE Environmental Conference Exploration and Production, San Antonio, Texas, March, 1993.

Sherwin, T.F., "Dealing with Naturally Occurring Radioactive Material in the Oil and Gas Industry," Petro-Safe '92, Conference Proceedings, Houston, Texas, January, 1992.

"Shoreham Deal Violates New York State Environmental Laws," *PR Newswire*, July, 1989.

Smith, A.L., "Radioactive Scale Formation," *SPE Journal*, June, 1987.

Spaite, P.W., and Smithson, G.R., *Final Report, Technical and Regulatory Issues Associated With Naturally Occurring Radioactive Materials (NORM) in the Gas and Oil Industry*, Gas Research Institute, Chicago, Illinois, April, 1992.

St. John, B., "Statewide Order 29-B and the Management of Oil Field Waste in Louisiana," Petro-Safe '92, Conference Proceedings, Houston, Texas , January, 1992.

St. Pe, K.M., "An Assessment of Produced Water Impacts to Low Energy Brackish Water Systems in Southeast Louisiana," Water Control Division, LDEQ, July, 1990.

Summerlin, J.Jr., and Prichard, H.M., "Radiological Health Implications of Lead[210] and Polonium[210] Accumulations in LPG Refineries," American Industrial Hygiene Association, *Journal*, Vol. 46, No. 5, 1985.

Suntrac Services, Inc., *Naturally Occurring Radioactive Materials Seminar Training Manual*, University of Houston, Institute for Environmental Management, 1992.

Taylor, W., "NORM in Produced Water Discharges in the Coastal Waters of Texas," SPE 25941, EPA/SPE Environmental Conference Exploration and Production, San Antonio, Texas, March, 1993.

Terrel, H.D., "What's Happening in Drilling," *World Oil*, V. 208, p.23, April, 1989.

Texas Department of Health, *The Texas Indoor Radon Survey*, April 20, 1992.

Thayer, E.C. and Racioppi, L.M., "Naturally Occurring Radioactive Materials: The Next Step," SPE, Global Responsibility: Proceedings of the First International Conference on Health, Safety and Environment in Oil and Gas Exploration and Production, Den Haag, Netherlands, November, 1991.

Toohey, R.E., *Radon vs. Lung Cancer: A New Study Weighs Risks*, Logos Argonne Natl. Laboratory, V.5, p.7-11, 1987.

Turner, R., "Radioactivity in Metal Scrap," NORM Ad Hoc Committee Meeting, Little Rock, Arkansas, February, 1991.

Wagner, J.F., "Toxicity and Radium[226] in Produced Water: Wyoming's Regulatory Approach," Wyoming Department of Environmental Quality, Proceedings, First International Symposium on Oil and Gas Exploration and Production and Waste Management Practices, New Orleans, Louisiana, September, 1990.

Waldram, I.M., "Natural Radioactive Scale: The Development of Safe Systems of Work," *Journal of Petroleum Technology*, August, 1988.

Willis, R.W., "1991 Legislative Strategy," *Petroleum Independent*, p.20, January, 1991.

Wilson, R. and Hendrix, W.G., "NORM: Occurrence, Handling and Regulation," Proceedings, Hazardous Materials Control Resources Institute, New Orleans, Louisiana, February, 1992.

Wilson, W. F., "NORM, Naturally Occurring Radioactive Materials," *Training Manual*, Strata Environmental Services, Inc., 1992.

ACRONYMS, SUMMARY FACTS, REPORTS, AND DEFINITIONS

COMMONLY USED ACRONYMS

Activity	The quantity of radionuclides described by the number of nuclear transformations occurring per unit of time.
AEA	Atomic Energy Act
AEC	Atomic Energy Commission
ALARA	As Low As Readily Achievable
ALI	Annual Limit of Intake
API	American Petroleum Institute
BEIR	Biological Effects of Ionizing Radiation
Bq	Becquerel
BRC	Below Regulatory Concern
BWR	Boiling Water Reactor
CERCLA	Conservation Environmental Response Compensation and Liability Act of 1980 (Superfund Act)
CFR	Code of Federal Regulations
Ci	Curie
CPSC	Consumer Product Safety Commission
CRPCD	Conference of Radiation Control Program Directors
CWA	Clean Water Act
DAW	Dry Active Wastes
DEC	Department of Environmental Conservation
DER	Department of Environmental Restoration
DHHS	Department of Health and Human Services

DOE	Department of Energy
DOT	Department of Transportation
EIS	Environmental Impact Statement
EPA	Environmental Protection Agency
FDA	Food and Drug Administration
FEIS	Final Environmental Impact Statement
FIFRA	Federal Insecticide, Fungicide and Rodenticide Act
FR	Federal Register
GAO	General Accounting Office
GCD	Greater Confined Disposal
HHS	Department of Health and Human Services (formerly HEW)
HLRW	High-Level Radioactive Waste
IAEA	International Atomic Energy Authority
Joule	The unit for work and energy, equal to one newton expended along a distance of one meter ($1J = 1N \times 1m$)
LDEQ	Louisiana Department of Environmental Quality
mrem	mrem millirem
NCRP	National Council on Radiation Protection
NEPA	National Environmental Policy Act
NIMBY	Not in my back yard
NORM	Naturally Occurring Radioactive Material
NRC	Nuclear Regulatory Commission
OSHA	Occupational Safety and Health Administration
ppb	Parts per billion

PPE	Personal Protection Equipment
ppm	Parts per million
PWS	Public Water Supply
RBE	Relative Biological Effectiveness
RCRA	Resource Conservation and Recovery Act
rem	Roentgen Equivalent-Man
SDWA	Safe Drinking Water Act
SI	Systeme international d'Unites
TRC	Texas Railroad Commission
TRU	Transuranic Radiation Units (elements or wastes having an atomic number > 92)
TSCA	Toxic Substances Control Act
TSSC	EPA's Toxic Substances Strategy Committee
TWC	Texas Water Commission
USC	United States Code
USGS	United States Geological Survey
VLLW	Very Low Level Radioactive Waste
VR	Volume Reduction

SUMMARY FACTS
SELECTED RADIOACTIVE ELEMENTS

RADIUM

1. The Latin derivation is "radius" or ray.

2. Atomic weight: 226

3. Atomic number: 88

4. Melting point: 700° C

5. Boiling point: 1140° C

6. Specific gravity: 5

7. Valence: 2

8.   Discovered in 1898 by M. and Mme. Curie in the pitchblende or uranite of North Bohemia. There is about 1 g of radium in 7 tons of pitchblende. The element was isolated in 1911 by Mme. Curie and Debirene. The pure metal is a brilliant white when freshly prepared, but blackens upon exposure to air.

9. Emits alpha, beta, and gamma radiation.

10. Radium^{226} emits radiation at 3.7×10^{10} disintegrations per second, which is equal to 1 curie with a half-life of 1602 years.

11. 25 isotopes of radium are now known to exist.

12. Radium loses about 1% of its activity in 25 years being transformed into other

elements of lower atomic weight. Stable lead is the final product.

13. Radium should be stored in a well-ventilated area to prevent the buildup of radon.

14. Inhalation, injection, or body exposure to radium can cause cancer and other problems. The maximum permissible limit for the whole body for Ra^{226} exposure is 7400 becquerel.

RADON

1. Its earlier name, nition, was derived from the Latin term "nitens" or shining.

2. Atomic weight: 222

3. Atomic number: 86

4. Melting point: -71° C

5. Boiling point: -61.7° C

6. Density of gas: 9.73 g/l

7. Specific gravity of liquid: 4.4

8. Specific gravity of solid at -62° C: 4.0

9. Valence: 0

10. Radon was discovered in 1900 by Dorn, who called it radium emanation. In 1908 it was named "nition" by Ramsay and Gray. They determined it was the heaviest known gas. Since 1923 it has been designated as radon.

11. 26 isotopes of radon are presently known.

12. $Radon^{222}$ is derived from radium. It has a half-life of 3.823 days and emits alpha radiation.

13. It is estimated that every square mile of soil to a depth of 6 in. contains about 1

g of radium, which releases radon to the atmosphere. Radon is present at the Hot Springs Resort in Arkansas.

14. At ordinary temperatures radon is a colorless gas. When radon is cooled below the freezing point, it exhibits a brilliant phosphorescence, which becomes yellow as the temperature is lowered and orange-red at the temperature of liquid air. Phosphorescent radon has been reported from oil-associated brine production in north central Texas at night in winter months. Radon clathrates have been described.

15. Radon is available at a cost of about $4 per mCi.

16. Radon is estimated to have caused many deaths in the United States. The maximum permissible buildup in confined spaces is 4 pCi/liter.

LEAD

1. Derived from the Anglo-Saxon "lead."

2. Atomic weight: 207.2

3. Atomic number: 82

4. Melting point: 327.502° C

5. Boiling point: 1750° C

6. Valence: 2 or 4

7. Long known metal; mentioned in *Exodus*.

8. Found in nature in the elemental state, but is rare. Most often found within the mineral galena (PbS).

9. Three lead isotopes are the stable remnants of radioactive decay. They are listed below:

(i)　　Uranium Series → Pb^{206}

(ii)　Actinium Series → Pb^{207}

(iii) Thorium Series → Pb^{208}

10. 27 other lead isotopes are radioactive, including Pb^{210}, which is commonly found in the equipment of gas plants.

11. Pb^{210} emits alpha, beta, and gamma radiation, with a half-life of 21 years.

12. Lead is a known cumulative poison, creating certain types of brain damage. Radioactive lead can cause cancer and cell destruction.

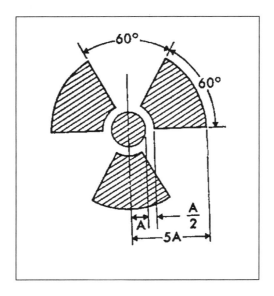

RADIATION SYMBOL

29 CFR 1910.96

1.   Cross-hatched area is to be magenta or purple.

2.   Background is to be yellow.

(2)   Radiation area Each radiation area shall be conspicuously posted with a sign

        or signs bearing the radiation caution symbol and the words:

CAUTION

RADIATION AREA

PERIODIC TABLE OF THE ELEMENTS

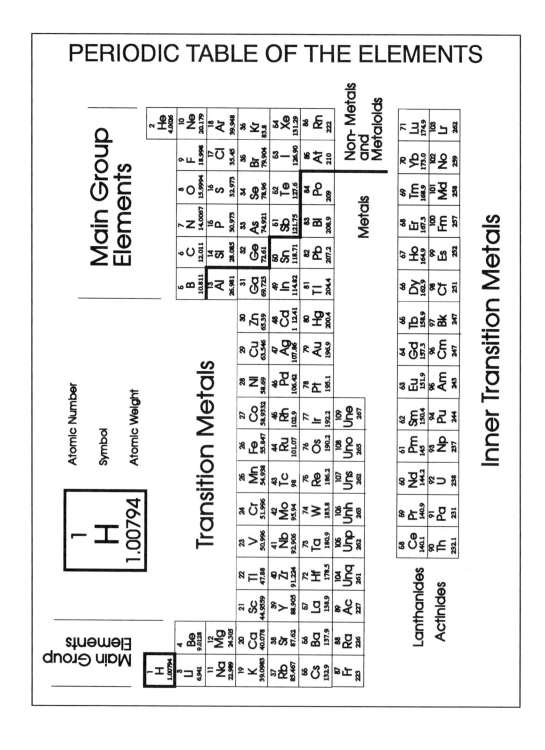

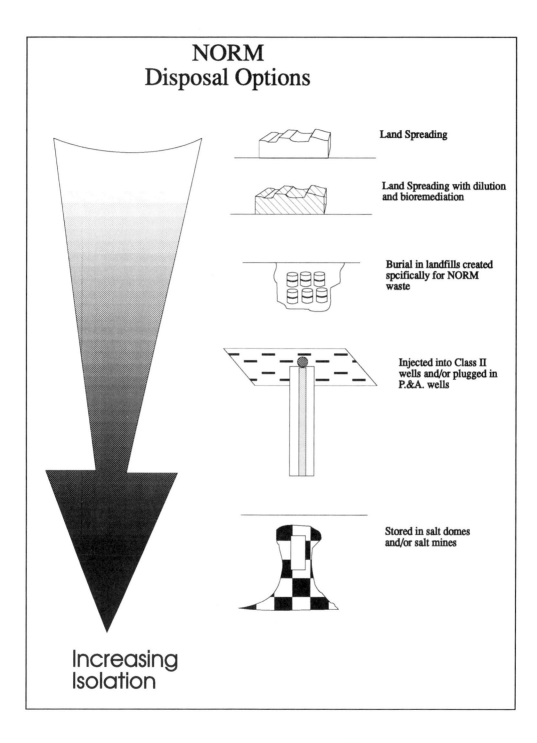

NORM
Disposal Options

Land Spreading

Land Spreading with dilution and bioremediation

Burial in landfills created spcifically for NORM waste

Injected into Class II wells and/or plugged in P.&A. wells

Stored in salt domes and/or salt mines

Increasing Isolation

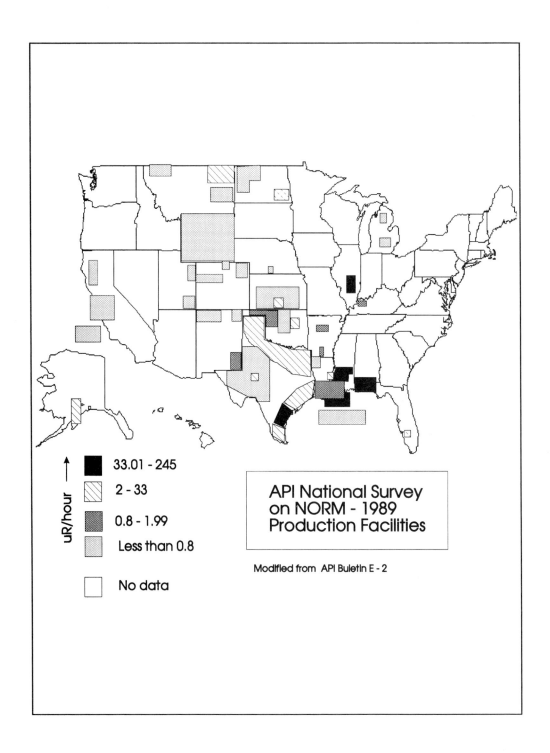

uR/hour →

33.01 - 245

2 - 33

0.8 - 1.99

Less than 0.8

No data

API National Survey
on NORM - 1989
Production Facilities

Modified from API Buletin E - 2

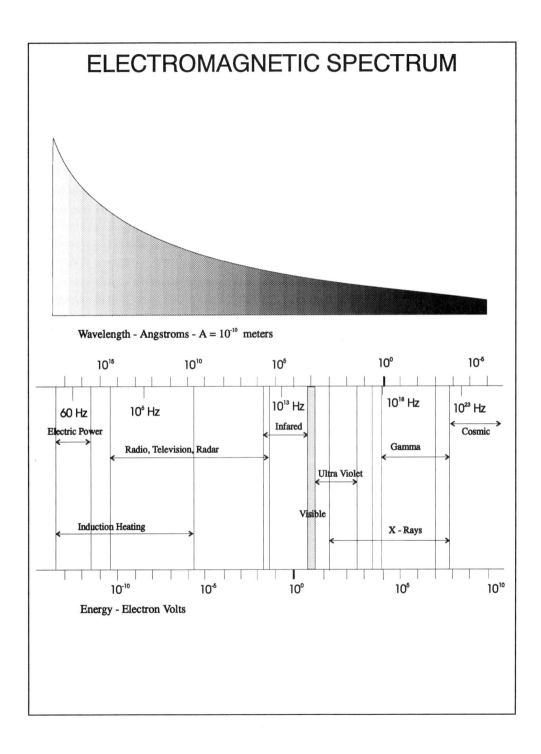

ELECTROMAGNETIC SPECTRUM

Wavelength - Angstroms - A = 10^{-10} meters

10^{15} 10^{10} 10^{5} 10^{0} 10^{-5}

60 Hz 10^{6} Hz 10^{13} Hz 10^{18} Hz 10^{23} Hz

Electric Power

Infared

Cosmic

Radio, Television, Radar

Gamma

Ultra Violet

Visible

Induction Heating

X - Rays

10^{-10} 10^{-5} 10^{0} 10^{5} 10^{10}

Energy - Electron Volts

INDEX

acronyms 196

Actinium Series 40, 42

activity level 33

Agreement States 5

airborne radon 60

Alabama 3, 125

Alaska 4, 125

alpha 11, 13, 21, 126, 129, 157

alpha emission 11, 14

American Petroleum Institute 6, 165

Angstrom 68

API 6, 165

Appalachian Mountains 53

aquifer 58, 63

Arkansas 3

atom 11, 13

Atomic Energy Act 124, 132

atomic number 16

atomic weight 16

background 73, 75, 126, 128, 135

barite 50, 83, 84

barium sulfate 50, 83, 169

Basin and Range 54, 61

Beaumont formation 75

bedrock 59

beta 11

Beta emission 11

Beta energy 14

Beta radiation 140

bibliography 183

biological effects 69

Black Hills 3

blood 107, 109

bone marrow 94, 105, 107, 109

breasts 74

brine 84, 92

brine well 167

bubble chambers 19

California 4, 87

Canada 5

cancer 45, 91, 94, 99, 113

carbon 26, 100

carbonates 170

cell 100, 102, 105

CERCLA 133

Chattanooga 53

chlorine 16, 26

chromosomes 102

circulatory system 114

Class I injection wells 167

Class II injection wells 126, 167

Clean Air Act 133

cloud chambers 19

CNS death 107

coal 87

coal ash 87

Code of Federal Regulations 130, 150

coefficient of risk 94

Colorado 4

Columbia Plateau 54

compliance 131

Conference of Radiation Program
    Control Directors 5, 134

Connecticut 44

consumer products 92

contour maps 178

conversions 76

cosmic radiation 56

counts per minute 73, 140

curie 69

cytoplasm 102

decay 11, 22

Department of Transportation 148,
    151, 172

deuterium 16

digestive tract 110

disintegrations per minute 72

disintegrations per second 69

disposal 126, 167

DNA 92

dosage 74, 77, 96, 107, 133, 142, 150

dosimeter 158

drinking water 129

electromagnetic spectrum 207

electron 12, 14

electroscope 158

Environmental Protection Agency 2 7, 86

ethane 173

evaporates 49

exposure calculations 97, 170

exposure damage 107

exposure limits 74, 75

exposure pathways 93

exposure rate 74

eyes 73, 112

Federal Register 124

Federal Regulations/Laws 132

fertilizer 86

field measurements 78, 164

Florida 4, 86

fuel 61

gall bladder 113

gamma 12, 14, 18

gamma radiation 14, 56, 140, 142, 164

gas 43

gas plants 169, 171

Gas Research Institute 6

Geiger-Muller 19, 156

genes 103

geology 49, 173

Georgia 44

geothermal waste 87

germ cells 105, 111

GI death 107

gonads 74

granite 64

Great Salt Lake 52

grid spacing 174

ground water 63, 85, 94

Gulf Coast 85

Gulf Coastal Plain 53

Gulf of Mexico 75

Hadron 17

hair 114

half-life 2, 19

Harris county 75

hazardous waste well 167

health risks 91

helium 14

high level waste 132

homes 58

hot spots 176

hot springs 87

Houston 3

hydrogen 15

igneous 49

Illinois 4, 125

illmenite 85

Impacts-BRC 97

Indiana 4

industrial sources 81

ingestion 93

inhalation 93

instrumentation 155

Inverse Square Law 142

involuntary exposure 100

isotope 15

joule 73

K^{40} 40, 45, 49, 56, 97

Kansas 4, 125

Kentucky 3, 125

kidneys 114

laboratory measurements 78, 164

Lake Agassiz 52

landfill 92

LD_{50} 107, 111

leachate 83

lead 169, 201

life span 115

lignite 87

Lissie Formation 75

lithium 15

liver 113

Louisiana 3, 75, 127

Louisiana Department of
   Environmental Quality (LDEQ)
   126, 175

lungs 113

lymph nodes 110

lymphatic system 110

lymphoid tissues 105

Marie Curie 68

measurement 67, 155

medical radiation 56

metamorphic rock 49

methane 84

Michigan 4, 125

Micro Roentgens/hr 73

midwest 85

Minerals Management Service 126

minors 76

Mississippi 3

mitosis 103

Montana 4

monzanite 85

muscle 105

mutations 103

NARM 134

natural gas 84

Nebraska Sand Hills 52

nervous system 111

neutron 12, 17

Nevada 54

New Albany Shale 53

New Jersey 5

New Mexico 3

Newark Series 44

NIOSH 146

non-occupational exposure 76

NORM vi, vii, 1, 39

North Dakota 4

North Sea 5

nuclear force 17

nuclear fuel 132

Nuclear Regulatory Commission 5, 97, 132

nucleons 12

nucleus 12

occupational exposure 74

Ohio 4, 125

oil 7, 43

Oklahoma 3, 125

OSHA 7, 71, 150

Part N 134

PATHRAE 97

pathways 93

Pb^{210} 169

Pennsylvania 4

Periodic Table of the Elements 15, 26, 204

permeability 56

Personal Protective Equipment 146

petroleum 83, 168

petroleum production equipment 164, 168, 171

phosphates 86

phosphogypsum 86

photon 14

plasma 107

platelet 107

Pleistocene 52

polonium 31, 172

positron emission 12

potassium 45

Practical Quantitation Levels 129

primacy 125

propane 174

protactinium 23, 31

protection 139, 146, 149

proton 12

Quality Factor 72

rad 72

radiation control area 174

radiation symbol 203

radiation units 68

radioactive decay 11

radioactive series 41

radioactivity 16, 18

radionuclides 2, 28, 33, 40, 63

radio-sensitivity 105, 146

radium 43, 64, 169, 199

radon 44, 54, 64, 169, 200

radon detector 160

radon maps 60

rare earth 85

Reading Prong 53

red cells 107

red phosphorus 85

regulations 123, 151

Relative Biological Factor 72

repositories 61

reproductive organs 105, 111

respirators 145, 149

Rocky Mountains 13

roentgen 70, 72

roentgen equivalent in man (REM) 71

roentgen per minute 72

Safe Drinking Water Act 125

salt deposits 61

sampling 163, 173

scale 168, 170

scientific notation 25

scintillation 157

scrap metal 166

sea water 50

sedimentary rock 49

Sierra Nevada 54

silicates 83

skin 114

slag 86

sludge 83, 169

soils 51, 59

somatic effects 105

South Carolina 44

spleen 110

state regulation 133

statistical grids 174

sterility 111

Superfund Act 133

surveys 164, 173

swamps 54, 178

tantalite 85

Technology Enhanced Natural
    Radioactive Material (TENR) 2

Texas 3, 119, 125, 133, 135

Texas Department of Health 135

Texas Natural Resource Conservation
    Commission 126, 135

Texas Railroad Commission 135

Texas Water Commission 126, 135

thorium 21, 42, 63, 132

Thorium Series 42

threshold 104

thyroid 74, 112

thyroxine 112

Time, Distance, Shielding 145

topography 178

toxicity 147

transmutation 21

transportation 151

transuranic waste 132

tritium 16

turbulent flow 164

United States Geological Survey 60

units of measurement 67

uranium 51, 63, 132

Uranium Series 24, 41

Utah 61, 167

vapor phase 95

voluntary exposure 100

Washington 5

waste 61

water treatment plant 85

water wells 58

weaponry waste 61

West Virginia 4

wetlands 178

white cells 107

Wilhelm Konrad Roentgen 72

Working Levels 76

Wyoming 4

X rays 70

Yucca Mountain site 61

zircon 85